FORESTS OF NEPAL

FORESTS OF
NEPAL

J. D. A. Stainton

HAFNER PUBLISHING COMPANY
New York
1972

In the U.S.A.
Hafner Publishing Company
866 Third Avenue
New York, N.Y. 10022

Text printed in Great Britain by
The Camelot Press, London and Southampton
Colour illustrations by
N.V. Grafische Industrie,
Haarlem, Holland

Contents

CONTENTS

MAPS

TABLES

Illustrations

ILLUSTRATIONS

NOTE

All of the following photographs were taken in Nepal. No. 106 was taken by Mr D. Sayers, the remainder by the author.

Most of the illustrations have been selected to assist people to recognise different species of trees and shrubs, rather than to show their habit of growth. For recognition close-up pictures are essential, and it will be evident that in many cases such pictures have only been obtainable by breaking off flowering branches and posing them at ground level.

1 Sal forest at 1,000 ft in the Rapti dun valley

2 Land cleared for agriculture in the Rapti dun valley, with sal forest on the foothills

3 The first stage in clearing forest for agriculture. Sal trees, ringed but still in leaf, at 1,500 ft in the lower Bheri valley. Wheat has been planted beneath them

4 *Castanopsis indica* near Pokhara in the Central Midlands. The lake is at about 3,000 ft

5 The road from the plains to Kathmandu at about 7,000 ft. The oak forest has here been reduced to a low scrub. *Rhododendron arboreum* in the foreground

6 *Pinus roxburghii* forest. The lack of under-shrubs is due to frequent fires

7 *Abies spectabilis* forest at 11,000 ft to the south of the main Himalayan range near the Rupina La

8 *Rhododendron arboreum* and deciduous forest at 10,500 ft on Lamjung Himal

9 Moist alpine scrub at 13,000 ft. *Rhododendron setosum* in flower

10 *Pinus excelsa* by the Rara lake at 10,000 ft, near Jumla

11 A village with flat-topped roofs near Jumla. *Picea smithiana* predominates in the forest in the background

12 Low bushes of *Caragana brevifolia* and *Lonicera spinosa* dominate the Dolpo landscape at 15,000 ft

13 In the Langtang valley *Betula utilis* forest grows at up to 14,000 ft

14 40-ft trees of *Juniperus recurva* at 13,000 ft near Thyangboche in Khumbu

15 Dry alpine scrub in which *Juniperus wallichiana* predominates, at 14,500 ft in Khumbu

16 *Michelia kisopa*

17 *Mahonia napaulensis*

18 *Michelia doltsopa*

19 *Camellia kissi*

20 *Schima wallichii*

21 *Stachyurus himalaicus*

22 *Bombax malabaricum*

23 *Ilex dipyrena*

24 *Skimmia laureola*

25 *Acer pectinatum*

26 *Buchanania latifolia*

27 *Sarauja napaulensis*

28 *Aesculus indica*

29 *Euonymus tingens*

30 *Butea frondosa*

31 *Butea frondosa*

32 *Butea minor*

33 *Acacia catechu* growing on river side gravel

34 *Caragana brevispina*

35 *Bauhinia purpurea*

36 *Albizzia mollis*

37 *Mimosa rubicaulis*

38 *Cassia laevigata*

39 *Caesalpinia decapetala*

40 *Erythrina stricta*

41 *Piptanthus nepalensis*

42 *Erythrina arborescens*

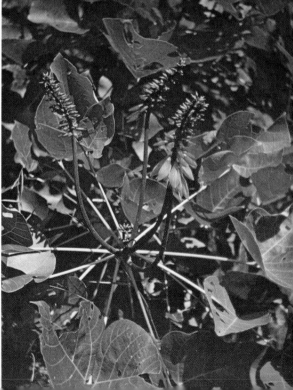

43 *Sorbus foliolosa*

44 *Prunus rufa*

45 *Prunus carmesina*

46 *Sorbus cuspidata*

47 *Spiraea arcuata*

48 *Spiraea bella*

49 *Rosa macrophylla*

50 *Photinia integrifolia*

51 *Spiraea canescens*

52 *Eriobotrya elliptica*

53 *Rosa sericea*   54 *Malus sikkimensis*   55 *Pyracantha crenulata*

56 *Rosa brunonii*   57 *Deutzia staminea*

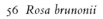

58 *Ribes griffithii*

59 *Hydrangea heteromalla*

61 *Ribes glaciale*

60 *Terminalia chebula*

62 *Dichroa febrifuga*

63 *Melastoma normale*

64 *Osbeckia stellata*

65 *Woodfordia fruticosa*

66 *Duabanga sonneratioides*

67 *Alangium salviifolium*

68 *Pentapanax leschenaultii*

69 *Brassaiopsis glomerulata*

70 *Torricellia tiliifolia*

71 *Cornus macrophylla*

72 *Cornus capitata*

73 *Helwingia himalaica*

74 *Lonicera hispida*

76 *Lonicera rupicola*

75 *Lonicera quinquelocularis*

77 *Zizyphus jujuba*

78 *Viburnum cordifolium*

79 *Viburnum grandiflorum*

81 *Viburnum cotinifolium*

80 *Viburnum erubescens*

82 *Abelia triflora*

83 *Eugenia frondosa*

84 *Luculia gratissima*

86 *Engelhardtia spicata*

85 *Dobinea vulgaris*

87 *Callicarpa arborea*

88 *Inula cappa*

89 *Pavetta indica*

90 *Microglossa albescens*

91 *Notonia grandiflora*

92 *Gaultheria hookeri*

93 *Enkianthus deflexus*

94 *Pieris formosa*

95 *Agapetes serpens*

96 *Vaccinium retusum*

97 *Gaultheria pyrolifolia*

98 *Vaccinium nummularia*

99 *Rhododendron cowanianum*

100 *Rhododendron lowndesii*

101 *Rhododendron triflorum*

102 *Rhododendron ciliatum*

103 *Rhododendron lindleyi*

104 *Rhododendron nivale*

105 *Rhododendron setosum*   106 *Rhododendron setosum* in the Langtang Valley

107 *Clematis montana*   108 *Rhododendron hodgsonii*

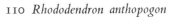

109 *Rhododendron cinnabarinum*   110 *Rhododendron anthopogon*

111 *Symplocos crataegoides*

112 *Ardisia macrocarpa*

113 *Symplocos sumuntia*

114 *Osmanthus suavis*

115 *Beaumontia grandiflora*

116 *Jasminum officinale*

117 *Jasminum dispermum*   118 *Jasminum pubescens*   119 *Ehretia wallichiana*

120 *Clerodendrum colebrookeanum*   121 *Clerodendrum infortunatum* by the Rapti river

122 *Phlogacanthus thyrsiflorus*   123 *Colquhounia coccinea*   124 *Leucosceptrum canum*

125 *Elsholtzia fruticosa*   126 *Aristolochia griffithii*

127 *Dodecadenia grandiflora*   128 *Litsea oblonga*   129 *Lindera pulcherrima*

130 *Litsea kingii*   131 *Neolitsea umbrosa* var. *consimilis*

132 *Daphne papyracea*

133 *Daphne bholua* var. *glacialis*

135 *Wikstroemia canescens*

134 *Daphne bholua*

136 *Scurrula elata*

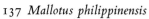

137 *Mallotus philippinensis*   138 *Euphorbia royleana*

139 *Quercus lanuginosa*(r.) and *Quercus incana* (l.)   140 *Quercus lamellosa*

141 *Lithocarpus pachyphylla*   142 *Quercus semecarpifolia*

143 *Castanopsis tribuloides*

144 *Castanopsis hystrix* (l.) and *C. tribuloides* (r.)

145 *Betula utilis*

146 *Betula alnoides*

147 *Salix sikkimensis* and *Rhododendron campanulatum*   148 *Salix calyculata*

149 *Juniperus wallichiana*

150 *Ephedra gerardiana*
     and *Hippophae*
     thibetana

151 *Pandanus furcatus*

152 *Podocarpus neriifolius*

153 *Cycas pectinata*

154 Crossing the Madi khola near Pokhara

155 Camp at the Rara lake, near Jumla

156 Porters on the march in the outer foothills, Central Nepal

Introduction

Although Nepal remained almost completely closed to western travellers until 1949 its flora before that date was by no means entirely unknown. Due to the efforts of early collectors, of whom F. Buchanan Hamilton and N. Wallich were the earliest and the most important, knowledge of plants growing in the vicinity of the Nepal valley and on the route up from the Indian plains was quite considerable. There had also been some smaller collections made from places high in the main mountain ranges. Knowledge of the flora of Nepal as a whole, however, was still sufficiently incomplete when her frontiers finally were opened for there to be a strong desire in British botanical circles to make further collections from a wider range of country. The flora both of Sikkim and of the Western Himalaya were well known by then, and although those parts of the Eastern Himalaya which lie to the east of Sikkim had been much less visited the collections of W. Griffith, F. Ludlow, G. Sherriff, F. K. Ward and others had yielded considerable information on the flora there. The intervening 500 miles of the Nepal Himalaya therefore constituted something of a blank in botanical knowledge of the Himalaya as a whole.

The first two British collectors to avail themselves of the new opportunities travelled as members of mountaineering parties. In 1949 O. Polunin visited Langtang, Rasua Garhi, and Chilime khola to the north of Kathmandu, and in 1950 D. G. Lowndes went further west to the Marsiandi valley and Manang. Their collections, which went to the herbarium of the British Museum (Nat. Hist.) and which included the new species *Rhododendron cowanianum* and *Rhododendron lowndesii*, whetted the British appetite for further botanical collecting.[1]

In 1952 the first expedition devoted entirely to botanical collecting and jointly sponsored by the Museum and by the Royal Horticultural Society visited the Humla–Jumla and Dolpo areas of West Nepal. The members of this expedition were O. Polunin, W. Sykes, and L. H. J. Williams. This was followed in 1954 by another expedition sponsored by the same institutions of which the members were J. D. A. Stainton, W. Sykes, and L. H. J. Williams. They collected in the country to the south of Annapurna and Dhaulagiri, also south-wards to Butwal and northwards to Mustang.[2]

These two big expeditions, together with the previous ones of Polunin and Lowndes, brought back a total of over 17,500 gatherings of plants for the herbarium of the Museum. With the resources at the disposal of the Museum it was obvious that this material would take a long time to work out, and despite the fact that these collectors had covered only part of the country official enthusiasm for further collecting in Nepal began to wane. In fact the number of gatherings from Nepal in the herbarium of the Museum now considerably exceeds the above figure, for with the collections made prior to 1949 and those made by myself and others in recent years the total number is probably around 40,000. This very large accumulation of material is still being worked upon, and it is hoped in the future to publish a list of all the species recorded from Nepal.

After 1954, therefore, there were no more big sponsored British botanical expeditions to Nepal. I myself, however, had received my initiation both to plant collecting and to Nepal during the expedition of that year, and as a result of my experiences in this very attractive country I was determined to return and collect further in unvisited parts. At this point I should explain that although I have collected on many occasions for the herbarium of the Museum and have received much assistance from the staff there I am not connected with the Museum in any official capacity.

I returned to Nepal for a six-month season of collecting in 1956, visiting the Arun and Tamur valleys in East Nepal. After this trip there was an interval of some years during which I travelled and collected in other parts of the world, and I did not return again to Nepal until 1962. Then during the eight years 1962–9 I spent part of each year in Nepal, travelling widely in this period through many different parts of the country. On a number of occasions when trying to trace the movements of collectors of the past I myself have experienced great difficulties due to lack of records; I therefore set out details of my own journeys in Nepal in a separate appendix for the possible convenience of others in the future.

The purpose of my journeys changed considerably over the years. In 1954 and 1956 my primary purpose was to amass as large and comprehensive a collection of plants as possible in order to strengthen the collections which the Museum already possessed. The search for specimens was conducted with something of the intensity of a treasure hunt, and the noting up and drying of the big collections, particularly during the wet and difficult conditions of

the monsoon months, absorbed almost all one's time. On my later journeys the necessity for making these big general collections was past, for by
then the Museum possessed representative collections from a wide range of
country. I was therefore able to be much more selective in my collecting, and
in consequence had far more time to devote to observations on the vegetation as a whole.

As time went by and my knowledge of the individual species grew I used to
find myself during the course of the day's march trying to bring some order in
my mind to the profusion of vegetation through which I passed. At first my
ecological observations were hardly more than a guessing game played to
amuse myself in the somewhat lonely conditions in which I travelled. I would
guess what the predominant vegetation was likely to be as I crossed to a slope
with a different aspect or climbed to a slope above; or on entering a certain
type of forest I would guess at which old friends among the trees I knew by
name I was likely to find within it. My next step was to begin to make notes on
the composition of the various types of vegetation, until on my later trips, and
fortified by now with knowledge of the standard ecological works published
on the Himalayan flora, the making of these notes became my primary objective. On these later journeys the collecting of herbarium specimens had become
only a secondary objective carried out very largely in order to be able to check
identifications of species in my ecological lists.

The following notes on the forests of Nepal are based in large part on those
lists. Being entirely self-taught in matters of botany and ecology I am well
aware that the notes must have many imperfections. I put them forward in the
hope that they may be of some assistance in providing a general picture of the
tree and shrub vegetation of the country, and in the belief that it may be a long
time before any properly trained botanist has either the opportunity or the
inclination to wander so widely through the hills of Nepal as I have done.[3]

In the matter of plant identifications I have received much help from the
staff of the Museum, and in particular from Mr L. H. J. Williams, but such
errors as there may be remain my own. An amateur botanist is ill-equipped to
pick his way through the shifting sands of botanical nomenclature. Since my
object is to make myself intelligible to those most likely to read these notes I
have in most cases retained the names used in the *Flora of British India*, but I must
confess that I have not been entirely consistent about this, and occasionally I
have used more recent synonyms.

Until the reader is familiar with, at any rate, some of the species mentioned

3

in these notes the descriptions of the various forest types will not have much meaning for him. The fundamental difficulty facing anyone who wishes to get to know the flora of Nepal is how to identify the individual species. The illustrated floras which provide a short-cut for beginners in many parts of the world are lacking here, and though there have been some books published in Japan with photographic illustrations of some of the plants which occur in Nepal they are by no means comprehensive.[4] I hope that the illustrations included in this book may be of some assistance in making identifications, but they also are far from comprehensive. Ultimately the only sure method of obtaining identifications is to collect specimens in the field and subsequently identify them in the herbarium. The seven stout volumes of the *Flora of British India* are too ponderous for field use, and in any case many of the species which occur in Nepal are not included in them. My advice to any beginner is to try to persuade someone with knowledge of the flora to accompany him on his first few marches and to impart information on the plants as they walk along.

Of necessity a botanical collector must spend much of his time in the field during the monsoon months, and there are moments when one's enjoyment of the country is somewhat dampened by the weather. These notes written in London have enabled me to revisit in my imagination some of the finest country in the world under the ideal conditions which by no means always prevail there. My debt of gratitude to all the men who accompanied me on these journeys is very great, for by their efforts on my behalf I have been able to visit much remote country. My especial thanks are due to Nima Sundar and the men of the villages around Chaunrikharka who came with me for many seasons. Their cheerfulness at all times added greatly to the pleasure of my journeys, and their resourcefulness under difficulties was immense. Without the steadfast assistance of men such as these a botanist is not likely to achieve very much in the roadless hills of Nepal

Climate

Rainfall figures for Nepal are scanty. Some that I have been shown by courtesy of the Hydrological Survey Department do not seem to be entirely reliable, for in many cases these figures have been recorded at stations which are not under the control of the Department. Figures which I quote for Nepal therefore should be accepted only with some reservation.

Any map purporting to show amounts of rainfall over the whole country should be treated with great caution. In this very mountainous country a generalised map, even if there were reliable records from a sufficient number of stations to justify its compilation, would have very little practical application. Amounts of rainfall vary sharply over very short distances due to local topography, so that records compiled at any station may have little relevance to the rainfall received at places close by.

General climatic factors

Nepal has a monsoon climate, with a pronounced rainfall maximum during the months of June to September. In nearby Darjeeling, out of a total annual average rainfall of 122 in. (3,100 mm), 102 in. (2,592 mm) fall during these four months.[1] The monsoon rains approach from the Bay of Bengal and reach the eastern end of the country first, so that the rains here begin earlier and end later than in the west, and the total rainfall in general is heavier. During the monsoon period rainfall in Nepal, though by no means continuous, is very frequent, and apart from very occasional brief moments early in the morning the big mountains are hidden in the clouds.

Rainfall figures recorded in the hills are often distorted by local topography, and figures from the plains give a clearer picture of the gradual diminution in monsoon rainfall as one moves westwards. Calcutta has an annual average of 59 in. (1,500 mm), Allahabad 38 in. (960 mm), Delhi 26 in. (660 mm), and Lahore 18 in. (457 mm).[2]

All parts of Nepal also receive a certain amount of rain during the winter months, and although this rain is only a small proportion of the total annual rainfall it is of considerable importance in that it enables a second annual crop to be grown. During the winter only a small part of the agricultural land of Nepal is irrigated, so that most of the cereal crops which are planted in the autumn and ripen in the early summer before the arrival of the monsoon are dependent on this winter rainfall. The rain varies from year to year, both in amount and in the time of its coming. In a year of good winter rains one sees good crops of wheat and barley ripening in the spring; in a poor year the crops on irrigated hillsides will still be good, but those without irrigation will be very poor.

In seeking reasons for this winter rainfall one must remember that the western end of the Himalaya has a rather different weather system from the eastern end. Cyclonic disturbances from Persia, and probably from Iraq and the Mediterranean, affect the western Himalaya during the early months of the year, giving the country there two rainfall peaks. In Malakand in North Pakistan out of an annual average rainfall of 37·3 in. (940 mm) 14 in. (355 mm) fall during July and August and 16 in. (406 mm) from January to April.[3] These disturbances bring heavy snowfalls to Kashmir, and their effect is felt through the Punjab and Kumaon to West Nepal. Their effect is diminished in eastern parts of Nepal, though there is still quite a considerable amount of winter rain here which on high ground falls as snow.

In a normal year the winter weather after the ending of the monsoon at the beginning of October can be summarised as follows. There is a very fine period until the end of November, when the snow mountains stand out in brilliant clarity in the rainwashed autumn sunlight. Then rather more broken weather often follows on to the New Year, though in some years rainfall at this time may be minimal. The early spring continues dry, broken by occasional brief periods of storms, but by this time the dust of the plains hangs at all but the highest altitudes and does much to obscure the distant views. The end of March brings the beginning of the hot weather, and with it an increase in thunderstorms; and from then to the advent of the monsoon at the beginning of June rainfall increases sharply due to these storms. Often the thunderstorms are very localised and their formation is in a large part due to topographical factors.

East Himalayan elements in the flora undoubtedly are much more abundant in areas of heavy rainfall in the eastern half of the country, but it seems probable that their presence there is due less to a high total annual rainfall than to a prolongation of the wet season by the early arrival and late departure of the monsoon. In many places where pre-monsoon thunderstorms are almost a daily occurrence from April onwards there is plenty of moisture in the soil for at least six months of the year, which for plants of the upper temperate zone covers the whole of the growing season. In contrast to this there are places in West Nepal where one can be knocking one's tent pegs into hard dry ground almost to the end of June.

On the tea gardens which lie at the base of the foothills on the north bank of the Brahmaputra in Assam, April and May are acknowledged to be wet months, and the rainfall which here precedes the arrival of the monsoon rains is too extensive to be explicable in terms of local thunderstorms. I do not know how far westwards in the Himalaya this pre-monsoon rainfall is generally supposed to extend. In the three seasons I have spent on the Arun and Tamur in the extreme east of Nepal the months of April and May have certainly been far wetter than I have experienced them to be in other parts of the country, with periods of continuous wet weather only in part supplemented by thunderstorms. I do not fully understand the reasons for the heavy pre-monsoon rainfall in Assam, but it seems reasonable to assume that much of the spring rainfall in the extreme east of Nepal is caused by the same forces which bring rain to Assam at the same time of year.

Southern sides of mountain ranges

The moist monsoon air which approaches Nepal from the plains is forced to rise when it meets the mountains. The increase in altitude leads to a decrease in temperature, and the clouds precipitate much of their moisture. In general, therefore, the southern sides of the mountain ranges receive a very heavy rainfall.

The country where the foothills meet the plains seems to have a substantially higher rainfall than places out in the plains. Biratnagar, for example, lies in the plains of East Nepal some 20 miles south of the foothills, and during the years 1949–60 received an annual average rainfall of 1687 mm. Dharan, lying north of Biratnagar at the immediate base of the foothills, received during this period an annual average of 2,366 mm.

Much of the rain which falls on the outer foothills both during the monsoon months and during the thunderstorms which precede them falls in downpour conditions. There are periods of intense rainfall, often quite brief, in which streams rise to fill their broad gravel beds and the country is awash with water, followed by spells of comparatively dry weather. The speed with which the

7

surface water clears from the sal forests is surprising, and the streams soon fall back to fordable proportions. Since rainfall during the monsoon must be ample even for the most moisture-loving plants one must assume that these downpours, though inflating the rainfall figures, do not greatly affect the vegetation. They are, however, the cause of much erosion in the outer foothills.

The country which lies at the base of the main Himalayan ranges is another area of very heavy rainfall. Pokhara lies close to the southern side of the Annapurna range, and here the monthly average rainfall in mm during the years 1957–65 was as follows:

Table 1   Rainfall at Pokhara

Jan	Feb	Mar	Apr	May	Jun	Jul	Aug	Sep	Oct	Nov	Dec	Total
35	30	49	82	174	654	807	925	515	170	22	14	3,477

Rainfall at Pokhara is higher than at any other place in Nepal at which records are kept. This is due in part to its proximity to the Annapurna range, and in part to the fact that the mountains which separate it from the plains to the south are low. In fact the rain gauge here is kept in the town, and the mountain slopes to the north must receive a considerably higher rainfall than does the town itself. Despite the very high rainfall of the Pokhara district I think that the distinction of being the wettest place in Nepal probably should be awarded to the country on the upper Arun in East Nepal.

The figures for Pokhara show clearly the rainfall peak which occurs there during the monsoon months of June to September, with the rain continuing on a little into October.

The considerable rainfall in May, and to a lesser extent in April, is the result of thunderstorms which build up almost daily during the hot weather. At that time of year thunderstorms are not uncommon everywhere in the hills, but they are especially frequent along the big southern faces of the main Himalayan ranges. Here every day as soon as the sun is well up one can watch the cumulus clouds forming, their white masses towering higher and higher to descend as rain later in the day. In fact the spring rainfall which results from these storms is very much heavier in places which lie immediately beneath the south-facing mountain walls than the figures recorded in the Pokhara valley would suggest. In late April on Lamjung Himal to the north of Pokhara I have been deluged with rain daily from midday onwards.

8

The rainscreen effect

The moist monsoon air, being forced upwards along the southern faces of the mountains, precipitates much of its moisture. When it passes over to the northern sides its moisture content is therefore much reduced, and the rainfall here is much lower. Where the rainscreen is formed by the main Himalayan ranges the contrast in rainfall between one side and the other is extreme, but contrasts in lesser degree can also be seen in many other parts of Nepal.

Jumla, which lies at 7,000 ft in West Nepal, illustrates very well the rain-screen effect. Like Pokhara it lies to the south of the main Himalayan snow ranges, but it lies much further into the hills than Pokhara and has a much lower rainfall. From Jumla a direct line to the plains is about 80 miles, compared to a distance of 45 miles for Pokhara. The monsoon air therefore has to cross a long stretch of mountain country before it reaches Jumla, and just before it arrives there it has to pass over a continuous chain of lekhs with a crestline of about 13,000 ft. In consequence the rainfall in Jumla is comparatively low. For the years 1957–64 the average monthly rainfall in mm was as follows:

Table 2   Rainfall at Jumla

Jan	Feb	Mar	Apr	May	Jun	Jul	Aug	Sep	Oct	Nov	Dec	Total
36	38	44	25	29	52	200	167	103	39	2	12	747

It can be seen that rainfall at Jumla during the monsoon months is only about one sixth of that received during this period at Pokhara. The low June figure is due to the fact that in western parts of Nepal the advent of the monsoon is often delayed until well on in that month. On the other hand rainfall from January to March is much the same in total amount as that received in Pokhara, and a much bigger proportion of the annual total. This latter point seems to reflect faintly the fact that at the western end of the Himalaya there is a second rainfall peak during these early months of the year.

Valley winds

In many of the big river valleys of Nepal a wind blows daily upstream, clearing away the clouds from the centre of the valley and sharply reducing the rainfall there. I discuss this phenomenon and its effect on the vegetation in greater detail in the section dealing with dry river valleys (*see* p. 37).

The gap through which the Kali Gandaki passes between Dhaulagiri and Annapurna provides an extreme example of reduction in rainfall caused by valley winds. The wind here is so strong that in the middle of the day it is something of an effort to walk against it. Jomsom lies at about 9,000 ft, and here even during the monsoon the centre of the valley is normally kept free of cloud, though the sides of the valley at that time of year are usually covered in mist. The up-valley wind here greatly reduces the rainfall, but it is still further reduced by the rainscreen effect, for by the time one has reached Jomsom one has passed to the northern side of the main ranges and is entering country where in general the rainfall is low. Both factors combine to give Jomsom the lowest recorded rainfall figures for any station in Nepal. For the years 1959–63 the monthly average rainfall in mm was as follows:

Table 3   Rainfall at Jomsom

Jan	Feb	Mar	Apr	May	Jun	Jul	Aug	Sep	Oct	Nov	Dec	Total
28	24	30	28	6	14	46	45	40	29	2	3	295

The monthly totals reflect in a somewhat accentuated form the same points we have already noticed at Jumla. The effects of the monsoon are hardly felt before July, and the disturbances coming from the west which bring rain and snow to the Western Himalaya during January to April bring a considerable part of the total annual rainfall to Jomsom.

Altitude

Heavy downpours which provide a considerable part of the total rainfall at lower altitudes become very much less frequent at higher altitudes. Here the rainfall tends to be less intense but more continuous, and during the monsoon on the southern side of the main ranges the upper slopes are almost perpetually covered with drizzling mist and cloud. The altitude at which the cloud hangs marks the upper limit of the zone of cultivation, for crops will not prosper in the dank and sunless air. This altitude is usually about 8,000 ft, though in very wet areas such as Annapurna Himal and the upper Arun cultivation ceases at around 6,000 ft, and in contrast in the Junbesi district where high ground to the south takes off some of the rain there are fields at up to 10,000 ft. In the Jumla and Dolpo districts and in the inner valleys conditions are altogether different, and cultivation may continue to over 14,000 ft.

Ascending during the monsoon through the dripping oaks and conifers, maples and rhododendrons which clothe the upper slopes on the south sides of the main ranges one is unlikely to find the alpine grazing grounds appreciably drier. Above the treeline, however, one is sometimes rewarded for a few moments early on monsoon mornings with a view to the south of peaks and spurs standing out like islands from a sea of white cloud which obscures all the lower ground and stretches away to the Indian border. These moments are brief, and the cloud soon sweeps up from the forest below to cover one in damp drizzling mist for the rest of the day.

If one climbs high enough in the alpine zone the actual rainfall is much diminished, for although the cloud cover during the monsoon remains almost constant it is only a thin Scotch mist with little precipitation. Most of the moisture content of the air has already fallen at lower altitudes. On the Khumbu glacier in the Everest region between April and November the low total precipitation figure of 390 mm has been recorded.[4] The Khumbu glacier lies in an inner valley with big mountains to the south of it, so that this low rainfall figure presumably results from the rainscreen effect at least as much as from altitude. I know of no rainfall records from high altitudes on the southern sides of the main ranges, but my observations of the flora lead me to believe that there also precipitation falls off significantly at great heights. I discuss this point at greater length in the section dealing with inner valleys (see p. 41).

Aspect

Aspect has a very important influence on the vegetation of Nepal. By looking at the various types of forests growing on the surrounding slopes it is often possible to orientate oneself as effectively as with a compass. South and east faces tend to be covered with very widespread species such as *Pinus roxburghii* and *Quercus lanuginosa*, whereas north and west ones have a more varied flora which is much richer in East Himalayan species. In general the steeper and shadier the face or gulley the more interesting the flora is likely to be.

This difference in vegetation must be due primarily to the fact that moisture is retained on the shady slopes for longer than on the sunny ones, and that the damp conditions favour the growth of certain seedling trees. An important secondary reason is that the undergrowth in the shady places is very much less combustible than in the sunny ones, so that the forest there is very much less modified by fire.

Aspect has a much more important influence on the vegetation at lower altitudes than at higher ones. Rainfall figures do not adequately reveal the constant high humidity of the upper zone which prevails throughout the whole of the comparatively short growing season. At lower altitudes the growing season lasts for much longer, and although the rainfall may be higher here the rate of evaporation is certainly much higher too, both because temperatures are higher and because there are far more intervals during the monsoon months when the sun shines.

Snow

Winter snowfall seems to vary considerably in amount from year to year, but undoubtedly it is heaviest in western parts of the country, and the winter snow-line here is lower than in the east. The snow also tends to lie longer into the spring in the west. In this respect it must be remembered that there is a difference in latitude of about four degrees between the two ends of the country.

The snow has cleared by the end of April from much of the alpine zone in eastern parts, but any collector who visits the snowless alpine slopes during May is likely to be disappointed, for there is nothing here in the nature of a snow flora to burst into flower amongst the melting snowdrifts. Most species wait to flower until well after the arrival of the monsoon, although there appears to be plenty of moisture in the soil during the intervening period.

By way of contrast it is worth quoting a statement of Sir Joseph Hooker and T. Thomson about the Western Himalaya.[5] 'As we penetrate further into the interior, a barren treeless climate rapidly succeeds, in which the principal vegetation occurs at the commencement of spring, when the melting snow supplies abundant moisture to small annual plants, which run their course with great rapidity and are speedily shrivelled up by the scorching sun.' The picture here presented is very different from that which one sees in the alpine zone of monsoon Nepal. Nor does it fit well with the vegetation of the dry Tibetan frontier areas of Nepal such as Dolpo.

I know of no rainfall figures for Dolpo, but the figures already quoted for Jomsom obviously have some relevance. One who has spent a winter in Dolpo informs me that snow falls occasionally throughout the winter, but that heavy falls are unusual and most of the snow disappears during the intervening periods of fine weather. Presumably much of the moisture content of this snow evaporates in the cold dry winter air without the snow ever melting. By the end of

May I have myself seen that there is no snow lying in Dolpo except in drifts on north faces. Certainly there is no true snow flora here, and annuals are not numerous. Most of the flora is composed of deep-rooted perennials, mats, and cushion plants, many of which do not come into flower until well on in July, months after the snow has gone.

PART I

Climatic & Vegetational Divisions of Nepal

Climatic & Vegetational Divisions of Nepal

In order to facilitate descriptions of the vegetation I divide Nepal into the following areas:

TERAI, BHABAR, DUN VALLEYS AND OUTER FOOTHILLS

Between the outermost foothills and the Indian frontier lies a belt of flat ground in places as much as 25 miles broad, which although geographically part of the Gangetic plain is politically part of Nepal. This country is known as the terai. At one time it must all have been covered with forest, but since the extensive flat ground here comprises some of the most valuable agricultural land in Nepal most of it is now under cultivation and treeless except for groups of mangoes and other cultivated trees around the villages. There are, however, a few parts of the terai, particularly in West Nepal, where quite large tracts of forest survive.

When seen from the terai through hot heavy haze the dim blue outlines of the foothills to the north appear to rise directly from the plains. In fact a closer approach almost always reveals that in the last few miles before one reaches their base the ground rises steadily for several hundred feet. This gently sloping country formed of alluvial gravels washed down from the foothills and accumulated at their base is known as the bhabar. Water is scarce here due to the porous nature of the soil, and streams which in the foothills may be sizable often disappear beneath the gravel on emerging into the bhabar. Many such streams run above the ground here only during the monsoon, and even then they fill their broad stony beds only briefly after periods of intensive downpour. As a result of this shortage of water villages in the bhabar are few, and some extensive forest survives. When seen from the air the bhabar often appears as a continuous belt of forest a few miles wide, with cleared terai land to the south of it and the foothills to the north.

The outermost foothills, which run in a line roughly parallel with the Indian border, are known as the Siwaliks, and in some parts of Nepal they are also referred to as the Churya hills. Their crestline does not much exceed 4,500 ft, and in many places is lower.

To the north of the Siwaliks and geologically distinct from them another and larger range known as the Mahabharat extends more or less continuously the whole length of Nepal. This range in places rises to over 9,000 ft, and lying at right angles to the course of all the rivers which flow down to the plains from the main snow ranges it channels almost all their waters into three great rivers, the Sapt Kosi, the Nayarani, and the Karnali. Where they debouch from their gorges into the plains these three great rivers are an impressive sight even during the dry weather; to stand on one of their banks during the monsoon and to see and hear the roaring destructive power of the dark racing waters is an awe-inspiring experience.

Within the outer foothills lie a number of broad gently sloping valleys which in the Indian hills are known as dun valleys. In Nepal they are also known as 'bhitri mardesh', which means 'the flat ground on the inside'. Although on the whole these valleys do not suffer from the water shortage of the bhabar much of the forest remains uncut here, and when the valuable flat lands of the terai are now largely under cultivation the question arises why the very similar flat lands of the dun valleys are still largely under forest. The principal reason seems to be that whereas the terai has always been open to pressure from people of Indian ethnic origin, the valleys lying within the hills have been under the control of the Nepali hill people who in the past have been reluctant to descend to these malarious levels. The almost treeless Dang valley in West Nepal, which appears to have been under intensive cultivation for centuries, owes its exceptional position in this respect to the predominance here of the Tharu people who are acclimatised to malarious conditions.

In some parts it seems also to have been the deliberate policy of the Government of Nepal to preserve an unpopulated zone against pressure from the plains. J. H. Burkhill[1] states that much of the Rapti valley to the south of Kathmandu was deliberately put out of cultivation by the Government at the beginning of the 19th century 'to build a barrier of malarious forest under the hills that no invading army should there obtain a base'.

Taken as a whole terai, bhabar, dun valleys, and outer foothills at the present time comprise much the most extensive forested areas of tropical and lower temperate Nepal. In recent years, however, pressure of population in the hills together with a considerable degree of success in the control of malaria have combined to draw the hillmen down to a lower altitude, and it is by no means impossible that in time to come these areas will be almost as completely deforested as most of the Midland areas are now.

The forests which grow here are composed predominantly of deciduous and semi-deciduous species. To anyone accustomed to medium altitudes in Nepal where evergreen species predominate it is somewhat surprising to find that in the spring at lower altitudes so many tropical trees are leafless. True tropical evergreen forest is very limited in extent here, being confined to shady gulleys and damp places. The seasonal monsoon rainfall does not permit the growth of tropical rainforest such as is found extensively in S.E. Asia.

In montane Nepal altitude and aspect are of paramount importance in determining the type of forest found at any particular place, but in terai, bhabar, and dun valleys this is not so. The ground here is more or less flat, so that the distinction between a north or south face which is so important elsewhere in the hills is not relevant here. Soil formation and the presence or absence of water in the subsoil replace the above mentioned factors as controllers of forest growth.

Forest in which sal (*Shorea robusta*) predominates is much the most extensive type of forest in all these flat places, and the composition of this forest varies remarkably little from east to west throughout Nepal. The sal will not grow on recently formed alluvium nor on waterlogged soil, so that along riversides it is often replaced by tropical deciduous riverain forest. In time the soil on these riverside terraces may mature sufficiently for the sal to succeed the original forest. Close to the river's edge on newly-formed gravels or on midstream islands there are often strips of forest composed of *Acacia catechu* or *Dalbergia sissoo*.

It is only along streams or rivers where riverain types of forest occur or in gulleys where tropical evergreen forest grows that the dominance of the sal forest is broken. Grassland is also a feature of some parts of these flat areas. In the tropical zone cultivated land when abandoned reverts very quickly to grassland, and in the Rapti valley I have seen fields which had been put out of cultivation to make way for a rhinoceros reserve only 18 months previously from which almost all traces of cultivation had already been obliterated by a dense growth of grass 6 ft tall (*Cymbopogon jwarancusa, Bothriochloa intermedia*). Presumably if left for long enough this type of grassland would revert to forest, but in some parts of the country grassland on ground which is waterlogged during the monsoon seems to be a permanent feature. In these places the grass may be 20 ft tall, and even when riding on the back of an elephant one can only just see over the top of this 'elephant grass'.

In the outer foothills there is a considerable difference between the types of forest found in eastern and western parts of the country, and I therefore describe the two ends of the country separately.

In the country north of Nepalganj and Dhangarhi in West Nepal most of the southern sides of the outer foothills at altitudes up to 4,000 ft are covered with subtropical deciduous hill forest, and the northern sides with sal forest. Above 3,500 ft there is much *Pinus roxburghii* forest and in some places this pine may descend as low as 2,000 ft. Above the pine belt *Quercus incana–Quercus lanuginosa* forest is widespread, though in one or two localities I have seen the oak meet the sal without the intervention of any pine. Tropical evergreen forest is extremely limited in extent and poor in species, occurring only in a few north-facing gulleys between 1–3,000 ft. The approximate sequence of forest types in the outer foothills of West Nepal is therefore as shown in Table 4.

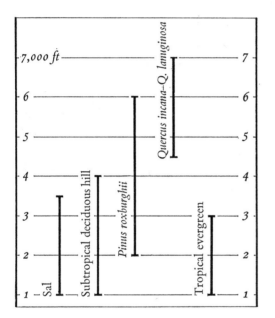

Table 4
Forest types on the outer foothills of W. Nepal north of Nepalganj

In the much wetter conditions prevailing in the foothills of East Nepal the forest sequence is different. Let us consider the foothills at the extreme eastern end of the country, between the Kosi and the Sikkim border.

Pinus roxburghii is rare here, and though it forms forest in some dry places in the Tamur valley I have not seen it do so on the outer foothills. I have not seen *Quercus incana* here at all, and *Quercus lanuginosa* forms forest only in a few south-facing sites between 4–6,000 ft. Above 7,000 ft there is a little *Quercus semecarpifolia*, but again only on south-facing slopes.

Both hill sal forest and subtropical deciduous hill forest are present in some quantity, but they have to compete with other types of forest which do not occur in West Nepal. *Schima-Castanopsis* forest grows between 2–6,000 ft, sometimes on south-facing slopes but much more commonly on north-facing ones. Tropical evergreen forest, though still confined to special sites such as gulleys and steep shady faces, is much more extensive here than in the west and is so much richer in species as to make one doubt whether it should be classed under the same forest type as the very poor western form. Subtropical evergreen forest of which *Eugenia tetragona* and *Ostodes paniculata* are perhaps the most typical component species occurs here between 3–5,500 ft in areas of heaviest rainfall. Michelias, laurels, and other species such as *Bucklandia populnea* form lower temperate mixed broadleaved forest between 5–7,000 ft, and in a few places between 6–7,000 ft there is forest composed predominantly of *Castanopsis tribuloides* and *Castanopsis hystrix*.

It will be seen therefore that in the extreme east of Nepal the western types of forest, although still present, have been replaced to a large extent by eastern types. The approximate sequence of forest types east of the Kosi is as shown in Table 5.

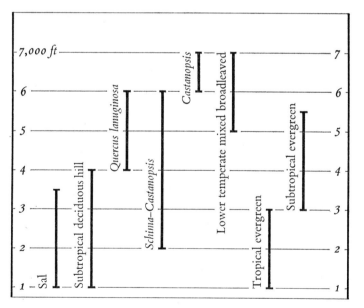

Table 5
Forest types on the outer foothills of E. Nepal east of the Kosi

THE MIDLAND AREAS, AND SOUTHERN SIDES OF THE MAIN HIMALAYAN RANGES

Between the outer foothills and the main snow ranges lie the Midland areas, which are densely populated. At lower altitudes most of the hillsides have been deforested and are now either terraced for cultivation or heavily grazed.

Most of the trees which grow around the fields and villages are planted to serve some purpose. Many of them produce something edible, such as *Myrica esculenta, Bassia butyracea, Tamarindus indica, Eugenia jambos, Ficus cunea, Moringa pterygospermum. Juglans regia* is principally grown in the drier western parts of the country, and here almost every village is surrounded by tall groves of walnut trees.

If any oak forest survives within reach of the villages it will be lopped heavily for fodder for the cattle, especially during the hot weather before the rains come. Many village trees are also used for the same purpose; *Sarauja nepalensis, Bauhinia variegata, Bauhinia purpurascens, Brassaiopsis hainla, Brassaiopsis aculeata, Celtis australis, Albizzia chinensis*, and *Grewia oppositifolia* are only a few of the very many species whose leaves are fed to cattle.

Trees are often planted in villages or at convenient places along the tracks to give shade to the traveller, and of these *Ficus religiosa* and *Ficus bengalensis* are the commonest. *Alnus nepalensis* is commonly planted for timber and to prevent erosion on steep wet slopes, and in eastern parts of the country *Cryptomeria japonica* is sometimes grown. *Prunus cerasoides* is often planted as a wayside tree.

Where any forest survives in the zone of cultivation its composition usually has been much altered by continuous lopping for firewood or fodder, or by burning the undergrowth to improve the grazing. Often the only remnants are coppices and shrubberies amongst which the cattle graze. The original forest at village level in many western parts of the country will have consisted of some form of oak forest, and here amongst the coppiced oak stumps one may commonly find any of the following shrubs:

Rosa brunonii, Prinsepia utilis, Pyrus pashia, Pyracantha crenulata, Deutzia staminea, Philadelphus coronarius, Spiraea canescens, Viburnum stellulatum, Coriaria nepalensis, Cornus macrophylla, Cornus capitata, Excoecaria acerifolia, Symplocos crataegoides, Toricellia tiliifolia, Rhus cotinus, Rhus wallichii, Syringa emodi, Abelia triflora, Jasminum humile, Jasminum officinale, Caryopteris wallichiana, Maesa chisia, Rhododendron arboreum, Lyonia ovalifolia,

Colquhounia coccinea, Inula cappa, Eurya acuminata, and species of Elaeagnus, Berberis, Rubus, Randia, Wendlandia, Maytenus, Indigofera, Ligustrum, Leptodermis, Buddleja, Zanthoxylum, and Elsholtzia.

The list is by no means a complete one, and many other shrub species can be found in these places.

In eastern parts of the country the slopes at village level will often originally have been covered with some form of *Schima-Castanopsis* forest, and here a few scattered trees of *Schima wallichii* or coppiced stumps of *Castanopsis* may survive amongst the shrubberies. Many of the species mentioned above as being common in shrubberies in western parts of the country are also common here, though certain of them such as *Rhus cotinus*, *Syringa emodi*, and *Abelia triflora* disappear under the wetter conditions, and others become a good deal less abundant. They are replaced by eastern species such as *Polygala arillata*, *Luculia gratissima*, *Viburnum erubescens*, *Dichroa febrifuga*, *Pieris formosa*, *Gaultheria fragrantissima*, *Camellia kissi*, and *Callicarpa macrophylla*. *Oxyspora paniculata* and species of *Osbeckia* also appear to be more common in eastern parts. Both *Daphne papyracea* and *Edgeworthia gardneri* owe their abundance in some of the village shrubberies here to their use in the past for paper-making.

Bamboos occupy an important place in the rural economy of Nepal. They serve a variety of purposes, such as poles for building houses, poles and ropes for building bridges, woven matting for roofs, piping for water supplies, containers for water or milk, trays for winnowing, and baskets for load carrying. Given a good supply of split canes the speed with which a man can weave himself a new basket is surprising.

At medium altitudes in the Midland areas of Nepal, where the country is intensively cultivated and very largely deforested, almost every family has its clump or two of bamboo planted somewhere around the house. Amongst the many uses to which these canes are put is the construction of big swings for the children during the autumn festivals. Usually situated on some open hilltop outside the village these tall constructions are a prominent feature of the country after the monsoon rains have finished.

It is unusual on the southern sides of the main ranges to find any permanent fields and villages above 8,000 ft, for above this height mist and cloud during the monsoon are too persistent to permit the satisfactory ripening of crops. In consequence much more forest survives between this height and the treeline at 12–13,000 ft than at lower altitudes, and though in some places the composition of this upper forest has been altered by burning, much undamaged forest still covers the remoter spurs and valleys.

In describing the vegetation of the Midlands and southern sides of the main ranges it is tempting to separate the generally well-preserved upper forest from the much-damaged remnants which occur in the zone of cultivation, but I refrain from doing so because although there is a very obvious manmade line at about 8,000 ft it does not seem to correspond with any significant natural division. In the few places where uncut forest survives in one continuous sweep from the subtropical zone to the temperate zone it is evident that the natural line dividing the two zones lies at around 6,000 ft. Since the country described extends the whole 500-mile length of Nepal it will, however, be convenient to divide it into an eastern, central, and western section.

The West Midlands

The country to which I refer in this section lies between the Kali Gandaki and the Kumaon border. To the north it is bounded by the Dhaulagiri range in the eastern part and by the Api-Saipal massif in the western part; in the centre its boundary is the line of lekhs which run to the south of Jumla rather than the main Himalayan crestline which here lies close to the Tibetan frontier.

The forest cover of the West Midlands is very similar to that described by A. E. Osmaston in his *Forest Flora for Kumaon*. Both on north and south faces *Pinus roxburghii* forest between 3–6,500 ft and *Quercus incana–Quercus lanuginosa* forest between 4,500–8,000 ft are very widespread, and so also is *Quercus semecarpifolia* forest above 8,000 ft. *Abies spectabilis* often succeeds the oak above 10,000 ft, but *Quercus semecarpifolia* may form an understory beneath the *Abies* and sometimes forms pure forest right up to the treeline. *Betula utilis* forest is widespread above 11,500 ft, and sometimes descends lower.

These are the forest types which occur most extensively in the West Midlands, but a number of other ones are also found here. *Quercus dilatata* forest occurs between 7–9,000 ft, usually being restricted to damp north faces. *Aesculus–Juglans–Acer* forest occurs between 6–9,000 ft along streams, and to a certain extent it overlaps *Quercus dilatata* forest in composition as well as in range of altitude. A somewhat impoverished form of lower temperate mixed broad-leaved forest containing species such as *Michelia kisopa*, *Lithocarpus spicata*, *Castanopsis tribuloides*, and *Machilus duthiei* extends as far west as the country between the Bheri and the Karnali, but I have not seen anything recognisable as this forest type west of the latter river. Rhododendrons, though few in

24

number of species compared to those found in the east of Nepal, are a common and numerous component of many types of forest found here, but I have not seen in this area anything resembling *Rhododendron* forest as I define it in the section on Forest Types (*see* p. 100).

Conifers are an important part of the upper forest. *Abies spectabilis* forest is widespread between 10,000 ft and the treeline. *Tsuga dumosa* forest is fairly widespread between 7–10,500 ft, and *Pinus excelsa* forest, most of which without a doubt is of secondary origin, occurs at much the same altitudes. I have seen *Cupressus torulosa* forest in the West Midlands only to the west of the Karnali, growing on south-facing limestone slopes between 7–9,500 ft. The low-altitude fir *Abies pindrow* I have likewise seen only to the west of the Karnali, growing between 7–9,500 ft on damp north faces; it is not as common here as it is in the Humla–Jumla area. Nor have I seen here two other conifers which are common in that area, *Cedrus deodara* and *Picea smithiana*.

Of the minor types of forest, small riverside woods in which *Alnus nepalensis*, *Populus ciliata*, or *Hippophae salicifolia* predominate occur at medium altitudes, and above the treeline moist alpine scrub grows on the alpine slopes at altitudes up to 14,500 ft. I have seen forest of *Juniperus wallichiana* only at Dhorpatan between 9,500–10,000 ft; its occurrence here to the south of Dhaulagiri must be due to exceptional climatic conditions, for in Nepal this species does not normally form forest to the south of the main Himalayan ranges.

Finally, some of the subtropical types of forest referred to in the section which deals with the outer foothills (*see* p. 19) penetrate up the big river valleys which cut deep into the Midland areas. Sal forest can be found here at up to 3,500 ft, and subtropical deciduous hill forest, often in a much damaged form, can be found up to 4,000 ft. Thickets of *Dalbergia sissoo* grow on the banks of the Karnali as far as 75 miles in a direct line from the plains, and *Acacia catechu* can also be found in places far into the hills.

The sequence of forest types in the West Midlands and on the southern sides of the main Himalayan ranges which lie to the north is thus approximately as in Table 6:

Table 6   Forest types in the W. Midlands

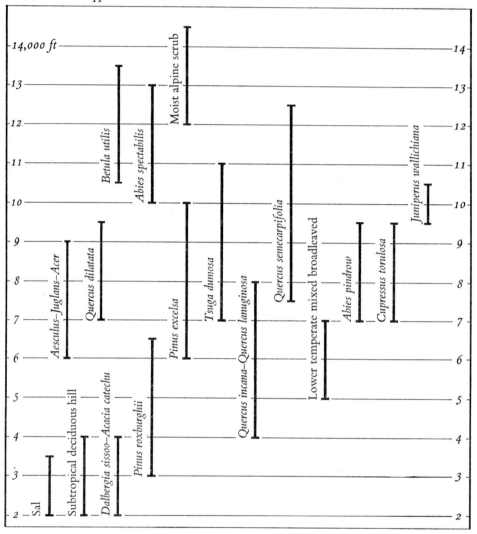

The East Midlands

The country to which I refer in this section is bounded to the east by the Sikkim border. I make the Arun–Kosi watershed the western boundary, because beyond this point there appears to be a quite appreciable reduction in Sikkimese elements in the flora.

The northern boundary is less clearcut. The big open valley of the Arun permits the passage of the monsoon rains right up to the Popti La, so that here the northern boundary is the Tibetan border. TheWalungchung, Yangma, and Kambachen valleys which lie at the head of the Tamur are more conveniently dealt with in the section on inner valleys, and so also are part of the country lying to the north of Topke Gola, and the head of the Barun.

Population is dense in the East Midlands, and the forests have been much cut at lower altitudes. Where they survive they show a similarity with those of Sikkim. In all the wettest places there is an almost complete absence of the types of forest which are widespread in West Nepal; for example, in much of the country on the upper Arun I have seen no *Pinus roxburghii, Pinus excelsa, Quercus incana, Quercus lanuginosa*, or *Quercus semecarpifolia*. They are replaced here by other East Himalayan forest types.

At subtropical altitudes *Schima–Castanopsis* forest often in a much damaged form is widespread, and subtropical semi-evergreen hill forest in which tree-ferns and *Pandanus furcatus* are prominent is common in wet shady places. They are succeeded between 5–7,000 ft by trees of the lower temperate mixed broadleaved forest such as *Lithocarpus spicata, Quercus glauca, Michelia doltsopa*, and many species of the family *Lauraceae*. In some places the forest between 6–9,500 ft may consist of a continuous belt of tall dark evergreen forest, the lower part composed predominantly of *Castanopsis tribuloides* and *Castanopsis hystrix*, the middle part of *Quercus lamellosa*, and the upper part of *Lithocarpus pachyphylla*. The *Quercus lamellosa* belt, however, is the only one of the three which is widespread, for the *Castanopsis* belt lies within the zone of cultivation and usually has been cut, and the *Lithocarpus* belt I have seen only east of the Tamur near the Sikkim border. More often the *Quercus lamellosa* belt is succeeded at about 8,500 ft either by the *Acer–Ilex–Magnolia–Osmanthus* association of the upper temperate mixed broadleaved forest, or in some parts by almost pure *Rhododendron* forest.

The predominance of *Rhododendron* species is a very marked feature of the upper forests of the East Midlands, and in some places the forest from 8,500 ft right up to the alpine shrubberies beyond the treeline consists almost entirely of trees and shrubs of this genus. *Abies spectabilis* and *Tsuga dumosa* may occur amongst the rhododendrons, but in the wettest places conifer forest is not extensive and the upper forest is predominantly broadleaved. Even *Betula utilis* in some parts seems to find difficulty in competing with the dense blanket of rhododendrons.

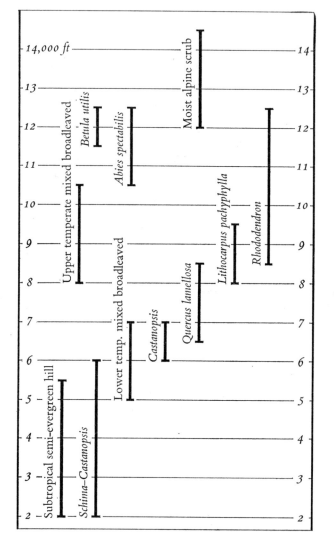

Table 7
Forest types in the wettest parts of the E. Midlands

Some of these East Himalayan types of forest extend into the Central Midlands, but there they are normally confined to north faces. In the wetter parts of the East Midlands not only are these types of forest much more extensive but they also often occur on south faces as well as north ones.

The vegetation of the less wet parts of the East Midlands, principally the big lower valleys of the Arun and Tamur, does not differ much from that of the Central Midlands; for example, *Pinus roxburghii*, *Pinus excelsa*, *Shorea robusta*, *Quercus lanuginosa*, and *Quercus semecarpifolia* can all be found there. Anyone wishing to study the sequence of forest types in these parts is advised to refer to the table for the Central Midlands on p. 31. For the wettest parts of the East Midlands the sequence is approximately as in Table 7.

The Central Midlands

(Except the country to the south of Annapurna and Himal Chuli.)

The country to which I refer in this section lies between the Arun–Kosi watershed in the east and the Kali Gandaki in the west. To the north I exclude the drier valleys which are dealt with in the section on inner valleys, the principal ones in Central Nepal being Khumbu, Rolwaling, Langtang, Shiar khola, upper Buri Gandaki, and upper Marsyandi. I also exclude the area to the south of Annapurna and Himal Chuli of which Pokhara is the focal point. The vegetation of this area is not typical of the Central Midlands, and I discuss it in a separate section.

Anyone who compares the sequence of forest types for the West Midlands with that for the wettest parts of the East Midlands will see that the forests of these two parts of the country have very little similarity. The West Midland forests resemble those of Kumaon, the East Midland forests those of Sikkim. The meeting of these different types of forest in the Central Midlands is complicated somewhat by the fact that the Pokhara area is exceptionally wet, but ignoring this point for the moment one can say that in most parts of the Central Midlands the western forest types are usually present on the south faces and the eastern forest types on the north faces.

Let us consider first the western forest types and see to what extent they occur in the Central Midlands. In the West Midlands the most widespread types of forest at intermediate altitudes both on north and south faces are *Pinus roxburghii* forest between 3–6,500 ft, *Quercus incana–Quercus lanuginosa* forest between 4–8,000 ft, and *Quercus semecarpifolia* forest between 8–10,000 ft. These types of forest are present in the Central Midlands also, but here usually they are restricted to south-facing slopes or the sides of big river valleys. At higher altitudes *Betula utilis* forest and *Abies spectabilis* forest are widespread in both areas, but extensive pure stands of *Abies* are a good deal more common in the Central than in the West Midlands because they are much less frequently broken up here by stands of high-altitude *Quercus semecarpifolia*.

Some types of forest, however, which occur in the West Midlands seem to be absent from the Central Midlands. I have not seen here *Quercus dilatata* forest, *Abies pindrow* forest, *Cupressus torulosa* forest, nor *Juniperus wallichiana* forest. The western *Aesculus–Juglans–Acer* forest is replaced by the *Acer–Magnolia–Osmanthus–Ilex* association of upper temperate mixed broadleaved forest. On damp ground, woods and plantations of *Alnus nepalensis* are common in the Central Midlands, but I have not seen here riverside thickets of *Populus ciliata* or *Hippophae salicifolia* such as occur in the west. Secondary forest of *Pinus excelsa* is not uncommon, but almost always it is restricted to south faces or drier river valley sites.

Tropical and subtropical forest is not extensive in the Central Midlands, and such as does occur there is usually much mutilated. Sal forest and subtropical deciduous hill forest grow in some of the lower valleys, and *Dalbergia sissoo* and *Acacia catechu* occur along the rivers in some places.

Now let us turn to consider to what extent the types of forest which are found in the wettest parts of the East Midlands occur also in the Central Midlands. *Schima–Castanopsis* forest below 6,000 ft and the *Michelia–Laurel–Lithocarpus* associations of the lower temperate mixed broadleaved forest between 5–7,000 ft are not uncommon in the Central Midlands, but they are very limited in extent and more or less confined to north or west faces. A few treeferns, *Pandanus*, and other trees of the subtropical semi-evergreen hill forest sometimes occur on damp shady faces, but I have not seen them forming true forest here. *Quercus lamellosa* is quite common on north faces between 7–8,500 ft, but frequently it occurs mixed with other oak species and not in pure stands as in the East Midlands. I have not seen *Castanopsis hystrix* here, nor any tall pure *Castanopsis* forest such as occurs in the East Midlands between 6–7,000 ft. *Lithocarpus pachyphylla* also appears to be absent.

In the upper forest *Tsuga dumosa* is dominant in some places between 7–11,000 ft, often with an undergrowth rich in rhododendrons. True *Rhododendron* forest in which this genus is dominant in the upper canopy occurs quite extensively in the Pokhara area, but elsewhere in the Central Midlands I have seen true *Rhododendron* forest only in insignificant amounts. The *Magnolia–Acer–Ilex* associations of the upper temperate mixed broadleaved forest on the other hand, though not extensive, are quite common on north and west faces.

The sequence of forest types in the Central Midlands excluding the country to the south of Annapurna and Himal Chuli is thus approximately as in Table 8.

Table 8   Forest types in the C. Midlands

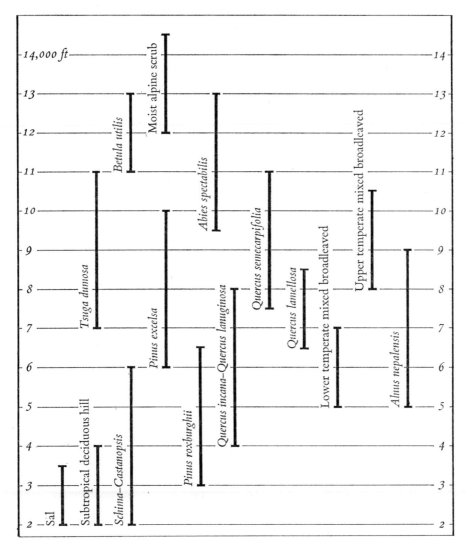

The Country to the south of Annapurna and Himal Chuli

In the section which deals with rainfall I point out that the country lying to the south of Annapurna and Himal Chuli appears to have a much higher rainfall than most other parts of the Central Midlands. The hills which lie between these mountains and the Indian border do not exceed 5,000 ft, so that during the monsoon the rainclouds make an unimpeded approach up the Madi and Marsyandi valleys to break with full force on the spurs of the big snow mountains. In most other parts of Nepal considerable areas of high ground intervene between the plains and the main Himalayan chain.

Even more important in its effect on the vegetation than the high summer rainfall is the fact that the wet season is prolonged by heavy thunderstorms during the spring. Premonsoon storms are a feature of all the country which lies immediately to the south of the main Himalayan ranges, but around Pokhara these storms seem to be specially frequent, probably because here a large area of low ground lies close beneath the high wall of mountains.

My experience of the Pokhara area is limited, and I cannot define precisely the extent of the country over which this high rainfall prevails. I have made two visits to Lamjung Himal and the head of the Madi khola, and the following account of the vegetation there will show that it differs considerably from that found in less wet parts of the Central Midlands. Due to the very persistent rainclouds which hang over the mountains here the villages and cultivation cease at about 6,500 ft, which is at least 1,000 ft lower than is usual in other parts of the Midlands.

At subtropical levels quite an amount of uncut forest survives. It consists mostly of *Schima–Castanopsis* forest, growing both on north and south faces. Treeferns, *Pandanus*, and other species of the subtropical semi-evergreen hill forest occur in many of the damp gulleys. Sal forest is not nearly so widespread as the *Schima–Castanopsis*, and is limited to dry south faces. *Pinus roxburghii* forest and subtropical deciduous hill forest, both of which one would expect to see at these levels elsewhere in the Central Midlands, are absent here.

At temperate levels *Quercus incana*, *Quercus lanuginosa*, and *Quercus semecarpifolia* also appear to be entirely absent; so also is *Pinus excelsa*. Between 5–7,000 ft there is often some lower temperate mixed broadleaved forest on north and west faces, with species such as *Michelia doltsopa* and many laurels giving to the forest an appearance very similar to that of the type as it occurs in the East Midlands. Between 6,500–8,500 ft *Quercus lamellosa* predominates both on north and south faces.

In the upper forest the heavy rainfall seems to drive out the conifers in the same way as it does in the wettest parts of the East Midlands. The forest is almost entirely broadleaved, and though *Abies spectabilis* is present it is limited in numbers and does not form extensive pure forest. I have seen no *Tsuga dumosa* on Lamjung Himal. Most of the forest consists of upper temperate mixed broadleaved forest or *Rhododendron* forest. It is interesting to note that although the growth of rhododendrons is very dense the forest species consist only of *R. arboreum*, *R. barbatum*, and *R. campanulatum*. Despite the fact that growing

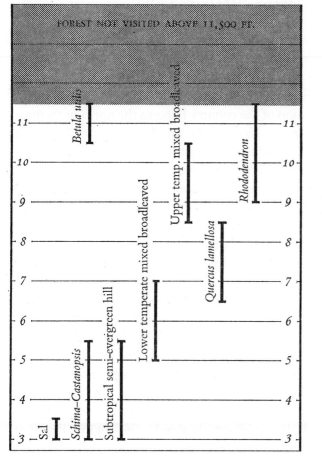

Table 9
Forest types to the S. of Annapurna and Himal Chuli

conditions look very suitable, species of the East Midlands such as *R. hodgsonii*, *R. grande*, *R. falconeri*, *R. thomsonii*, or *R. campylocarpum* have not been recorded here. *R. dalhousiae*, however, is present.

Certainly the Lamjung forests are not as rich in East Himalayan species as are those of the East Midlands, but in their general appearance they bear a closer resemblance to the forests of the East Midlands than do any other that I have seen in any of the country which intervenes between this point and the Arun–Kosi watershed. The forest sequence is approximately as in Table 9.

THE HUMLA–JUMLA AREA

This area is bounded to the south by the long chain of lekhs which lie north of Jajarkot and Dailekh and extend between the Bheri and Karnali rivers. The crest-line of these lekhs is between 13–15,000 ft, and by cutting off much of the rain which during the monsoon months comes up from the plains they modify very greatly the climate and vegetation of the country lying to the north of them.

To the west of Jumla the area is bounded by the Karnali river. This river divides in its upper part into the Humla and Mugu branches, and in the Humla Karnali valley the western boundary is formed by the snow mountains of which Saipal is the highest peak. To the north the Tibetan border is the boundary, and to the east it is formed by Kanjiroba Himal. The upper part of the Mugu Karnali valley is more conveniently dealt with in the section on dry inner valleys, and I suspect that the upper Humla Karnali valley beyond Simikot and other frontier valleys here which I have not visited might well also be referable to that section.

Jumla itself lies at about 7,000 ft, and anyone who visits the surrounding countryside will immediately be aware that he is in a land very different from the Midland areas of Nepal. The country is not nearly so deeply eroded as in these wetter parts, and most of the tracks are suitable for the ponies which are bred on the wide grass meadows. Clear streams run through gently sloping valleys, and the absence of large snow mountains gives to the landscape an unusually open appearance. Scenically the country has a certain resemblance to Kashmir, and in the section dealing with distribution I point out that the West Himalayan element in the flora is strongly represented here.

It is a country of great charm during the summer months. The streams are lined with chestnut and poplar; the fields surrounded with shrubberies of *Rosa, Deutzia, Syringa, Cornus, Cotoneaster, Viburnum,* and *Jasminum*; and the extensive mixed conifer woods which cover the hills are interspersed with damp meadows bright with summer flowers. Cultivation continues up to 10,500 ft, in contrast to the usual Midland limit at 8,000 ft, and even rice is grown at up to 9,000 ft. The flat roofs of all the village houses indicate a comparatively low rainfall.

North of Jumla lies the Rara lake, said to be the biggest in Nepal, set in a saucer of pine-clad hills. North again of the lake lies the Humla district, which is dominated by the big valley of the Humla Karnali. At the point where this river divides from the Mugu Karnali the altitude is 4,500 ft, and even as far upstream as Simikot the river is only at 7,500 ft, so that the centre of Humla is formed by a large area of low-altitude country. Most of the main valley has been cleared of forest, but in the side valleys where conditions are wetter some extensive forest survives very similar in composition to that of the Jumla area.

The forests of the Humla–Jumla area differ a good deal from those of the West Midlands which adjoin them to the south. The *Pinus roxburghii* and *Quercus incana–Quercus lanuginosa* forests which cover so much of the latter country at medium altitudes are not widespread in Humla–Jumla, where they are confined mostly to the Humla Karnali valley. Much of the Humla–Jumla area is covered with mixed conifer forests. The commonest species is *Pinus excelsa*, which forms extensive pure forest between 7–10,500 ft on south-facing slopes. The predominance here of this pine is certainly due in part to the destruction or modification of the original forest cover by man, for it is able to regenerate on disturbed sites very much faster than other conifer species. Pure stands of *Picea smithiana* are confined mostly to north-facing or west-facing slopes in less disturbed localities, but elsewhere individual trees of this species often occur mixed with the pine.

Tsuga dumosa is rather scarce over much of the area, and where it does occur it is often only a minor component of the *Pinus–Picea* forest. There is, however, some quite extensive *Tsuga* forest between 9–11,000 ft in one or two of the side valleys of the Humla Karnali.

Both *Cedrus deodara* and *Cupressus torulosa* are commonly grown in the area as village trees. *Cedrus* forest occurs between 6,500–8,000 ft in the Tila khola west of Jumla, and less extensively in the Humla Karnali below Simikot. *Cupressus* forest occurs in the Khater khola which leads from the Rara lake down to the Karnali.

In the Jumla area I have seen scattered trees of the low-altitude fir *Abies pindrow* only in a few localities, but in the Humla area this species forms tall forest between 7–10,000 ft on many north-facing and west-facing slopes. Sometimes it grows in pure stands, and sometimes mixed with *Picea smithiana*.

In all these conifer forests there are numerous streams and gulleys where the conifers give way to broadleaved trees. Strips of *Aesculus–Juglans–Acer* forest are common here. Of other minor types of forest found in damp places *Alnus nepalensis* woods are not as common as they are in the Midlands, this species to a large extent being replaced by *Populus ciliata*. *Hippophae salicifolia* also often grows along the streams.

Table 10   Forest types in the Humla–Jumla area

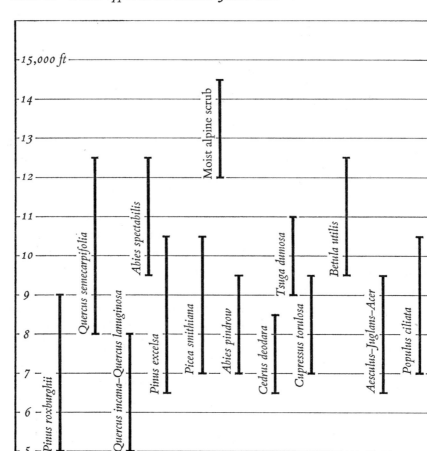

Above 10,000 ft *Abies spectabilis* is widespread, but unbroken tracts of pure *Abies* forest such as one sees extensively to the south of the main ranges in Central Nepal are not so common here, because the *Abies* tends to be broken up by stands of *Betula utilis* or *Quercus semecarpifolia*. The oak often forms pure forest on rocky south-facing slopes right up to the treeline. The birch is dominant on slopes and in gulleys where the snow lies late, and may descend as low as 9,500 ft.

Moist alpine scrub occurs above the treeline, but it is difficult to make a clear distinction in some parts of the area between this moist type and dry alpine scrub. For example, the juniper of the wet southern sides of the main ranges, *Juniperus recurva*, is here very largely replaced by the juniper of the dry country, *Juniperus wallichiana*.

The sequence of forest types in the Humla–Jumla area is approximately as in Table 10.

DRY RIVER VALLEYS

The rivers of Nepal cut deeply into the mountains, and the vegetation in their valleys is often so different from that of the immediately surrounding country-side that a separate section is required to describe it.

The bigger rivers in their lower courses, i.e. where their waters are below 3,000 ft, are usually broad and comparatively smooth flowing, and except during the rains there are many places where one can cross in safety in canoes hollowed out of big logs of *Bombax malabaricum*. The valleys are wide, and often there is old river-terracing on either bank. Here there will be an upstream wind during the day, but as there are no big mountain masses nearby the force of the wind usually is mild and it has no significant effect on the vegetation.

Further north, where their waters may be flowing at 4–5,000 ft, many of these rivers become roaring boulder-strewn torrents, confined between mountain walls which tower up for many thousands of feet on either side. Here the effect of the daily upstream wind is much more significant. If one visits such a place during the rains the valley floor is usually free of mist and rain, and perhaps even a small strip of blue sky may be visible overhead. The sides of the valley for 3,000 ft or so are usually clear also, but the slopes which lie above normally will be blanketed with mist and rain. It is obvious that as a result of the upstream wind rainfall in the centre of the valley and on the lower slopes is much reduced. It is equally obvious that this reduction in rainfall does not occur on the upper slopes.

In these places the last few hundred feet of the slopes down to the water are often very steep and sometimes precipitous, and since most of the bigger rivers run in a general north–south direction a steep eastern bank which faces west gets very little sunshine. Here therefore there may be a surprisingly damp habitat in the midst of a dry one, and this is especially so when the shadiness of the site is reinforced by a permanent supply of water emerging to the surface close to the riverside from the catchment of the slopes above. In consequence these big river valleys often provide very sharp contrasts in habitat within a short distance; damp sites close to the river; dry sites for several thousand feet above; and above this, slopes which are at least as wet as (and perhaps wetter than) than those of the surrounding country.

Whatever may have been the original vegetation of these valleys many of their slopes are now deforested by cutting and burning. Where forest survives it usually consists of *Pinus roxburghii* below 6,500 ft and *Pinus excelsa* above. This latter pine ascends to a very high altitude in some parts of Nepal, but in these valleys the belt of forest that it forms is usually a narrow one. Above 8,000 ft conditions on the slopes of many of these valleys become much wetter, and the pine is usually replaced by other types of broadleaved forest above this altitude.

The valley of the Bhote Kosi (also known as the Rongsha Chu) which flows down from Tibet past the Rolwaling valley and Charikot in Central Nepal provides a fairly typical example. At the big suspension bridge SE. of Charikot the river flows at about 3,000 ft, and at its confluence with the Rolwaling khola some twenty miles to the north it flows at 5,000 ft. The gorges of the Bhote Kosi are not on the same gigantic scale as those of the Kali Gandaki or Buri Gandaki, and much of the valley here is fairly open, but near the Rolwaling confluence it narrows between high grey cliffs. Over much of this 20-mile stretch there are steep banks along the riverside, and in these damp shady places species typical of wet subtropical forest occur, such as *Pandanus furcatus* and *Cedrela toona*, with very many epiphytic ferns and orchids and species of *Aeschynanthus*, *Hoya*, and *Raphidophora*.

Above this wet riverside strip the slopes are drier, and it is evident that many of them are burnt periodically. Some are covered only with grass, but in places there is a good deal of *Euphorbia royleana* growing on south-facing rocks. Whether the *Euphorbia* here is confined to the rocks as a result of fire or because of the dryness of the site I do not know. *Butea minor* also grows here, another species with a liking for dry rocky gorges in Nepal.

Pinus roxburghii grows in some quantity in the valley as far as the Rolwaling confluence. Forest even of this very fire-resistant pine cannot survive regular burning for ever, for though the big trees survive the fire the young seedlings are destroyed. In these valley sites one often sees tall mature pine trees which from the absence of any young trees obviously are destined to give way to grass slopes in the future. *Pinus excelsa* succeeds *Pinus roxburghii* at about 6,000 ft in some places here, but shortly above this height wet conditions begin again on the valley sides and the upper forest is of the wet broadleaved type. This 20-mile stretch of the Bhote Kosi valley exhibits therefore a typical wet-dry-wet sandwich effect.

Excluding the Tukucha valley on the upper Kali Gandaki which I deal with in the section on inner valleys (*see* p. 47), much the driest and windiest valleys I have seen in Nepal are those of the Karnali and the Bheri. Here the force of the wind during the day is sufficiently strong not only greatly to reduce the rainfall in the centre of the valley but also to have a direct dessicating effect on the vegetation, so that a markedly xerophytic flora is found.

Let us consider first the Karnali valley. Between the confluence of the Mugu and Humla branches of the river at 4,500 ft and its confluence with the Tila khola at 3,000 ft, the distance is about 50 miles. All this section of the valley is subjected during the day to a very strong upstream wind of a force sufficient to make at least one of the big rope bridges which span the river unsafe to use until the wind drops at nightfall.

The river banks at altitudes up to 4,000 ft are lined in many places with a thin strip of *Dalbergia sissoo*, the roots of which are fed with moisture from the river. This species is accompanied by some *Acacia catechu* and a few trees and shrubs such as *Olea cuspidata*, *Celtis australis*, *Zizyphus oxyphylla*, *Adhatoda vasica*.

The slopes above the riverside are treeless for a height of about 3,000 ft. It seems doubtful whether these windblasted slopes ever supported forest of any kind, but if they did so it has long ago been burnt off them. In many places, however, there is a kind of forest composed of *Euphorbia royleana*, the plants growing densely packed together to a height of 20–30 ft. This species occurs quite commonly in many of the big river valleys of Nepal, but I have not seen it elsewhere growing in anything like the profusion that it does on the Karnali. Many of the rocks are covered with the succulent and strangely leafless stems of species of *Sarcostemma*, which hang down 15 ft or more from overhanging rock-faces. On the rock ledges grow *Capparis spinosa*, *Withania coagulans*, *Pulicaria petiolaris*, and *Hyoscyamus muticus*.

The very dry conditions persist for about 3,000 ft above the valley floor. Above this height there are a number of fields and villages, and such forest as remains uncut appears to be of types more or less normal for the district at this altitude, i.e. *Pinus roxburghii*, *Quercus incana*, *Quercus lanuginosa*, *Quercus semecarpifolia*.

Now let us turn to the Bheri valley. Very dry conditions probably begin on the valley floor at about the point where the river passes round the western end of the Dhaulagiri massif near the isolated peak of Hiunchuli Patan, but here the river flows in a deep gorge and is inaccessible. Up at Tibrikot the valley is more open, and there is a riverside track along the very dry 10-mile section between this place and the Suli Gad confluence to the east. At Tibrikot the river is flowing at somewhat below 7,000 ft; and despite the spot-height recorded on the old survey map at the Suli Gad confluence I do not think the river at this point has risen by more than a few hundred feet.

In this part of the valley the wind during the day is very strong, and the vegetation on the lower slopes is reduced to a steppe-flora of grass and *Artemisia* more typical of the Middle East than of monsoon Nepal. The presence on the rocks of *Capparis spinosa* and a few blasted shrubs of pomegranate and fig add to the Middle Eastern resemblance. In one or two places there are oases of green amongst the prevailing greys and browns where irrigated fields are fringed with hedgerows and fruit trees, but for most of this 10-mile distance the only trees are a few scattered specimens about 15 ft tall of *Olea cuspidata*. This almost treeless steppe country continues to the Suli Gad confluence, where there are some stands of *Cupressus torulosa*. Beyond this point the wind still has an important effect on the vegetation of the valley, but it is more convenient to refer to the upper Bheri together with the very windy valley of the upper Kali Gandaki in the section dealing with inner valleys (*see* p. 41).

The dessicating effect of the wind does not extend so high up the valley sides of the Bheri as it does on the Karnali, and on north-facing slopes the forest begins about 2,000 ft above the valley floor. Most of the higher slopes are covered with *Pinus–Picea* forest of a similar kind to that which occurs widely in the Humla–Jumla area, but there is also an interesting little band of *Cedrus deodara* sandwiched in between the grass slopes and the *Pinus–Picea*. This band is only a few hundred feet in depth, but it runs almost continuously to the Suli Gad confluence and beyond. Most of the trees are stunted and very openly spaced, and it seems that the cedar is only just able to survive in these marginal conditions. This species here reaches the eastern limit of its range.

Taking the Karnali and Bheri valleys together there are a number of succulent plants occurring here which are well adapted to resist the dessicating effects of the wind: *Euphorbia royleana, Sarcostemma brevistigma, Sarcostemma brunonianum, Aloe indica, Kalanchoe spathulata, Notonia grandiflora,* and a species of *Caralluma.* Other species recorded from these valleys which are not at all typical of the flora of monsoon Nepal include *Olea cuspidata, Sophora mollis, Plectranthus rugosus, Sageretia thea* var. *bornmuelleri, Acacia farnesiana, Hyoscyamus muticus, Withania coagulans, Capparis spinosa, Onosma thomsonii, Pulicaria petiolaris, Pistachia integerrima.*

INNER VALLEYS

Rainfall during the monsoon in the valleys which lie deep within the main Himalayan ranges is significantly less than it is at similar altitudes on the southern sides of these ranges. In discussing the vegetation of these valleys it will be convenient to divide the country in half on the line of the Trisuli river to the north of Kathmandu, because in the western half of the country this diminution in rainfall begins at lower altitudes than it does in the eastern half. Much of the country marked on Map 1 as inner valleys in fact consists of snow mountains in which these valleys lie.

Inner valleys, from the Trisuli eastwards

The main inner valleys I have visited in the eastern part of the country are the Kambachen, Yangma, and Walungchung valleys at the head of the Tamur; the Thudam and Barun valleys on the Arun; Khumbu, Rolwaling, and Langtang. Undoubtedly there are a number of other valleys here, both big and small, which have a similarly reduced rainfall at their heads.

All these valleys in their lower parts have the steep water-worn outline typical of mountain areas of heavy rainfall. Above about 11,000 ft the profile of these valleys changes to the U-shape typical of ice-worn country. The valley floors become broad and open, and in some places the valley sides are formed by smooth ice-worn slabs of rock. Here the country offers to the traveller very much less laborious walking than do the knife-edge ridges and precipitous gorges of which so much of the mountain country of Nepal consists.

At the heads of most of the valleys are big glaciers, their terminal moraines lying at about 14,500 ft surrounded by the great snow-clad peaks of the Himalaya. The right time for a botanist to visit these places is in July, when most of the plants are in flower and most of the mountains are in the clouds; but although at that time of year drizzling mist and low cloud hang for much of the day just above the valley floor there are often moments early in the morning when one can catch a glimpse of huge ice-peaks towering up out of the mist.

These broad valleys provide extensive grazing grounds for yaks and sheep, and in some places there are herdsmen's huts as high as 16,500 ft. Many of the smaller valleys are inhabited only during the summer months, but of the main valleys named above all except the Barun are permanently inhabited. Most of the settlements lie between 11–13,000 ft, but there are also subsidiary villages lower down the valleys for winter grazing and spring cultivation.

No significant reduction in rainfall appears to take place in most of these eastern valleys until altitudes close to the treeline are reached. In consequence the composition of the forest here does not differ much from that of the forest which grows on the southern sides of the main ranges. *Larix*, however, which I have never seen on the southern sides, grows in the Kambachen, Yalung, and Langtang valleys, and *Pinus excelsa*, which is not a normal component of the upper forest in the wetter parts of Nepal, often grows quite extensively in these eastern inner valleys, though much of it is probably secondary in origin.

The reduction in rainfall at higher altitudes in the valley heads is marked by a change in the junipers. On the south sides of the main ranges the normal juniper species growing both as a tree in the upper forest and as a dwarf shrub in the alpine zone up to 14,500 ft is *Juniperus recurva*. The principal juniper of the dry areas of Nepal is the black juniper, *Juniperus wallichiana* (*J. indica*), growing both as a shrub and as a small tree. In the heads of the eastern inner valleys *Juniperus wallichiana* meets *Juniperus recurva* at about 12,000 ft, and by 14,500 ft the former species predominates.

As well as the black juniper which dots the hillsides with low round clumps at altitudes up to 15,500 ft other alpine shrubs indicate the change to drier conditions. Mats of a prostrate species *Myricaria* grow on riverside gravel; *Ephedra gerardiana* on stony terraces; and *Hippophae thibetana*, *Spiraea arcuata*, *Lonicera myrtillus*, and species of *Caragana* on the valley floor where the cattle graze. *Iris kumaonensis* occurs in these heavily grazed places, and a very big-flowered form of *Rosa macrophylla*, which seems to be confined to drier areas, often grows in the shrubberies round the villages. High on the alpine slopes *Rhododendron nivale* also indicates a reduced rainfall.

The herb flora also reflects the same influences. The following is a list of species either seen by myself in the eastern inner valley or recorded in the collections of others from these places. These species are all plants which are not usually found to the south of the main ranges:

Tanacetum gossypinum, Tanacetum nubigenum, Allardia glabra, Soroseris hookeriana, Artemisia biennis, Artemisia capillaris, Artemisia moorcroftiana, Saussurea gnaphalodes, Saussurea gossypifera, Saussurea graminifolia, Saussurea hieracioides, Saussurea hookeri, Saussurea leontodontoides, Saussurea sacra, Saussurea tridactyla, Cremanthodium decaisnei, Aster diplostephioides, Polygonum macrophyllum, Polygonum tortuosum, Meconopsis horridula, Gentiana algida var. przewalskii, Delphinium brunonianum, Selinum cortusoides, Phlomis rotata, Eriophyton wallichianum, Primula wollastonii, Primula buryana, Primula cave-ana, Androsace muscoidea, Androsace delavayi, Guldenstaedtia himalaica, Sibbaldia purpurea, Pinguicula alpina, Pyrola rotundifolia, Arisaema jacquemontii, Veronica lanuginosa, Oreosolen wattii, Pedicularis elwesii, Pedicularis sculleyana, Pedicul-aris longiflora var. tubiformis, Lancea tibetica.

This list certainly is not an exhaustive one, but it is sufficient to show that there is a large element in the herb flora here which is more typical of transhimalayan country than of monsoon Nepal.

Because many of these inner valleys run in an east–west direction it might be supposed that their heads are dry solely because they are protected from the monsoon rains by the high mountain walls which form their southern flanks, i.e. that they are directly comparable with transhimalayan country which is dry because it lies to the north of the main ranges. Although their dryness undoubtedly in part is due to this cause it cannot be the only operative factor, for a number of valleys which run in a north–south line show similar signs of reduced rainfall at their heads even though in their lower parts they offer an unimpeded approach to rainclouds coming up from the south.

Altitude obviously is an important factor, for rainclouds by the time they reach high altitudes have already deposited most of their moisture on the slopes beneath. Even on the southern sides of the main ranges rainfall diminishes at really high altitudes. It seems probable that at the heads of the inner valleys the moisture content of the rainclouds has already been much reduced by their passage up long valleys hemmed in by mountain walls, so that the altitude at which the clouds cease to precipitate rain and turn to drizzling mist lies some thousands of feet lower in these places than it does on the outer slopes.

Inner valleys, from the Trisuli westwards

I have visited only a few of the many inner valleys in the big stretch of country which lies between the Trisuli river and the Kumaon border. There are good descriptions of some of these valleys available. That curiously isolated little bit of the extreme west of Nepal, the Tinkar khola, has been described by E. Schmid from information supplied by Heim and Gannser.[2] It is not surprising to find that this valley, sheltered from the south by large mountain masses, is very dry. The Shiar khola, Buri Gandaki, and Marsyandi valleys have been described by J. Kawakita,[3] and from his account it is clear that these valleys are much drier than the southern sides of the main ranges.

I describe here the vegetation only of the upper Bheri and upper Kali Gandaki valleys. It is evident that these valleys share with those described by Schmid and Kawakita one characteristic which distinguishes them from the inner valleys of the eastern half of the country: diminution in rainfall begins at altitudes well below the treeline, so that it affects the composition of the forests growing there to a much greater extent than in the east.

The upper Bheri valley

The upper Bheri river has two main tributaries, the Suli Gad and the Tarap khola. The three valleys through which these rivers flow are sheltered from the rain to a considerable extent by the Dhaulagiri range to the south, and they provide habitats suitable for the drier types of forest.

Conifers predominate in most places. *Quercus semecarpifolia*, which in many parts of West Nepal is widespread right up to the treeline, is scarce in these valleys and appears to be altogether absent from their heads. Since elsewhere this species will grow as high as *Betula* or *Abies* it is tempting to assume that it is dryness rather than cold that prevents its growth here. This explanation, however, is not entirely convincing when one remembers that this oak in other places is often found growing in sites which are drier than average, such as south-facing rocky slopes. It may be relevant at this point to recall one rather surprising feature of the Kashmir forests; oak is absent from the conifer forests which lie to the north of the Pir Panjal range despite the apparent suitability for its growth of much of the terrain there. One plausible explanation given for its absence, which also might well be applicable to the heads of these Nepal valleys, is that the oak was driven out of the Kashmir valley during the comparatively recent ice age and has not as yet succeeded in recolonising the lost ground.

Of the conifers much the most widespread species from 7,000 ft right up to the treeline is *Pinus excelsa*. In hot sunshine the country is filled with the resinous smell of its branches. Most of the ground in these valleys is too high for *Pinus roxburghii*, though near Tarakot in dry windy conditions this pine grows to the unusually high altitude of 9,000 ft.

Forest of *Cupressus torulosa* is quite extensive, particularly in the Suli Gad at altitudes between 7–9,500 ft. It seems probable that in the past cypress forest has been more extensive in these valleys than it is now, and that in many places it has been replaced either by grass slopes or by *Pinus excelsa* as a result of burning.

In view of the fact that *Picea smithiana* is a species which prefers damp sites it is surprising how far up these drier valleys it manages to survive. There are stands of *Picea* a long way up the Suli Gad at altitudes up to 10,000 ft, and it also grows far up the upper Bheri, but it is absent from the highest and driest forest.

I have not noticed any *Tsuga dumosa* or *Abies pindrow* in these valleys. In the section on dry river valleys I have described the belt of *Cedrus deodara* which grows in the Bheri valley near Tibrikot. This belt runs more or less continuously up the Bheri as far as Tarakot. There are also some stands of cedar amongst *Pinus excelsa* and *Picea smithiana* in some of the side valleys here at altitudes up to 9,000 ft, but the cedar does not range very far from the main valley.

In the upper forest above 10,000 ft *Abies spectabilis* is often present, but in the driest parts it becomes scarce and to a large extent is replaced by *Pinus excelsa* and *Betula utilis*. In part this absence of the *Abies* may be due to burning, but it seems clear that at very high altitudes on the edge of the treeless Dolpo country the pine has a greater capacity for survival than the fir.

In the absence of oak, broadleaved forest is represented only by *Aesculus–Juglans–Acer* forest at lower altitudes and *Betula* forest above. The former type of forest is not at all extensive, but in the Suli Gad on riverside flats there are small woods between 9–10,000 ft consisting of *Ulmus*, *Acer*, *Populus*, *Juglans*, *Prunus*, and *Euonymus*.

Betula utilis is widespread in the upper forest. This species, together with *Pinus excelsa* and *Juniperus wallichiana*, appears to be the tree species best able to survive at high altitudes in dry conditions. The treeline here is usually much higher than in wetter parts of Nepal, and in many places there is extensive birch forest above 14,000 ft. The usual associate of the alpine birch in most other parts of the country, *Rhododendron campanulatum*, persists for a surprisingly long way up these valleys, but it is absent from the driest places.

Juniperus wallichiana is common at all altitudes in these valleys both as a tree and as a small shrub. In the drier parts it forms a kind of low forest, with trees scattered very openly on the rocky slopes. This forest is only barely distinguishable from the dry alpine scrub into which the juniper merges in dwarf alpine form. The scrub (which consists of species such as *Spiraea arcuata*, *Rosa sericea*, *Ceratostigma ulicinum*, *Ribes alpestre*, *Ephedra gerardiana*, and species of *Caragana*), although referred to as alpine, often descends without any very noticeable change in its composition as low as 10,000 ft. In many places the composition of this scrub has been much modified by grazing and burning.

The north-facing slopes of the Dhaulagiri range which form the south side of the upper Bheri valley appear to be wetter than the country described above, and the forest growing here is rather similar to that of the Jumla area. Ignoring

Table 11   *Forest types in the valleys of the upper Bheri and its tributaries*

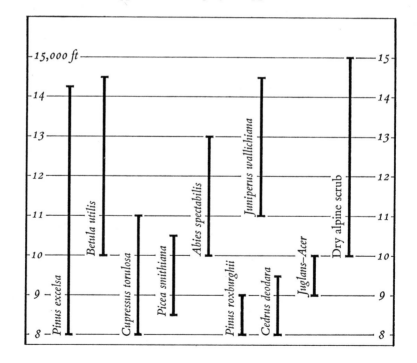

this wetter part the sequence of forest types in the upper Bheri valley and its tributaries the Suli Gad and the Tarap khola is approximately as in Table 11.

The upper Kali Gandaki valley

One of the principal charms of Nepal is the great variety of climate and vegetation contained within a comparatively small country. Nowhere in Nepal have I seen so abrupt a change of climate and vegetation telescoped into so short a distance as in the valley of the Kali Gandaki, where this river passes through the narrow gap between Dhaulagiri and Annapurna. In the space of a day's march one can descend from the sharp dry sunshine and windblown Tibetan flora around Jomsom to the steamy humidity and subtropical monsoon flora of the gorges below Ghasa.

The abruptness of the change is astonishing. All the inner valleys become progressively drier towards their heads, but here the transition is accelerated by the very strong wind which blasts its way up from the south through the gap in the mountains during most of the daylight hours, and which clears the rain-clouds from the centre of the valley. In fact the transition is very much less abrupt on the valley sides than it is on the valley floor, and forest grows in some of the side valleys here far to the north of the point at which the valley floor becomes treeless.

Let us start at Jomsom and follow the valley southwards towards the rain. Jomsom lies at about 10,000 ft, and here in the centre of the valley the flora is reduced to a few blasted shrubs of *Sophora morcroftiana*, *Lonicera hypoleuca*, and species of *Caragana*. The only trees are a few willows and fruit trees growing on irrigated land. *Nepeta leucophylla* grows in tufts on the river banks. The slopes for about 2,500 ft above the valley floor are also treeless and covered only with tight round bushes of *Caragana*. Between 12,000–13,500 ft there is some steppe forest of *Juniperus wallichiana* and a good deal of dry alpine scrub, and above this again there are alpine slopes which are rather wetter than any of the lower country. A little *Betula utilis* grows amongst the juniper, but I have seen no real birch forest here.

Moving down the valley towards Marpha the valley floor is still very dry and covered only with *Ephedra gerardiana* and species of *Caragana* and *Artemisia*, but the slopes above are much wetter and the forest on them consists of *Abies spectabilis*, *Betula utilis*, and *Pinus excelsa*. In side valleys there is also some *Picea smithiana*. Squeezed in between the dry valley floor and the wetter forest above is a narrow belt only a few hundred feet in depth of open steppe forest composed of *Juniperus wallichiana* and *Cupressus torulosa*.

Tukucha lies at about 8,500 ft, and here the valley broadens out into a wide expanse of gravel through which the river meanders. There are some much cut trees of *Cupressus torulosa* on the valley floor here, and it seems probable that at one time *Cupressus torulosa* forest extended further towards Jomsom. The valley sides continue to be covered with *Pinus excelsa*, *Abies spectabilis*, and *Betula utilis*. Broadleaved trees are still scarce here, and *Quercus semecarpifolia* appears to be absent.

Not much further change occurs until one reaches Kalopani some miles to the south. There is *Pinus excelsa* here, and some *Cupressus torulosa* as well, but many species which one does not see in the drier country to the north appear at this point, such as *Tsuga dumosa*, *Aesculus indica*, *Rhododendron arboreum*, and species of *Taxus*. A mile or two further on one reaches true monsoon country, sees unirrigated fields of maize and potatoes, smells the moist warm fragrance of monsoon vegetation, and begins to pick the leeches off one's legs. From here southwards the flora is typical of Midland Nepal.

If one does this journey in the reverse direction and sees the forest gradually peter out into the treeless steppe—country which leads up to Mustang, the immensity of Asia is very vividly impressed on one by the reflection that the nearest forests to the north lie about a thousand miles away in the Altai and Tien Shan across the treeless Tibetan plateau and the Takla Makan deserts. E. Schmid points out[4] that the main region of steppe forest lies in north temperate latitudes between the Eurasian conifer belt and the steppe zones, and that in comparison the steppe forest belt in the Himalaya is a very narrow one. Certainly this is so in the western half of Nepal, and in the eastern half of the country this belt seems to be altogether absent.

Schmid makes another point about the steppe forest belt. 'From the want of endemics we conclude a comparatively young age of that belt in the Himalayas.' In so far as this statement is limited to endemic species occurring exclusively in steppe forest it is no doubt true, but taking the inner valleys of the western half of Nepal as a whole the number of new and apparently endemic species recorded from them in recent years is rather surprising. *Primula sharmae*, *Primula ramzanae*, *Primula reidii* var. *williamsii*, *Primula poluninii*, *Rhododendron lowndesii*, *Clematis phlebantha*, *Clematis alternata*, *Micromeria nepalensis*, *Lamium tuberosum*, *Nepeta staintonii*, *Cremanthodium purpureifolium*, *Saussurea chrysotricha*, *Saussurea platyphylla* are only some of the new species recently recorded from this area.

Since prior to the opening of Nepal to botanists in 1949 almost no collecting had been done in these drier parts one would expect a certain number of new species to be found here, but now that the whole of Nepal has been fairly well collected it is somewhat surprising that in proportion to their size these areas have produced far more new and apparently endemic species than any other part of the country. It is true that these areas have been subjected to exceptionally intensive collecting both by the Japanese Kyoto Expeditions of 1952–3 and by the British Museum Expeditions of 1952–4, but I do not think that this fact alone can entirely explain the number of new species recorded.

THE ARID ZONE

North of Dhaulagiri and Annapurna Himal lies an almost treeless country which in climate and vegetation is Tibetan in character. This arid zone falls into three parts; Dolpo, Mustang, and Manang.

Manang lies at the head of the Marsyandi, sheltered from the rains by the whole length of the Annapurna range. The vegetation here has been described by J. Kawakita.[5] I have not visited this country, and I do not refer to it further.

Mustang I visited during the course of my first visit to Nepal, at a time when I had not learned to make adequate ecological notes on the vegetation of the country through which I passed. I therefore make no comment on the country here except to point out that even as far north as Mustang the wind which blows up through the Tukucha gap still controls to a certain extent the rainfall. The centre of the valley here obviously has a low rainfall, whereas the mountains on either side are much wetter. In July and August the valley sides above 14,500 ft are very constantly covered with mist, and they receive heavy rain at times. The valley centre, even at the height of the monsoon months, usually is dry and sunny. In consequence the vegetation occurring below 14,500 ft is of a very much drier type than that occurring above this altitude.

Dolpo

The name Dolpo, which does not appear on the survey map, is applied by the local Tibetan-speaking people to the country drained by the Barbung khola and the Langu khola. The watershed with the upper Kali Gandaki is the eastern boundary of the district, and the northern one is the Tibetan border. To south and west the boundaries are drawn so as to include Mukut, the Barbung khola, and Tarap, but to exclude Ringmo and Mugu.

All this country has a low rainfall, being sheltered from most of the effects of the monsoon by the Dhaulagiri range to the south. The northern parts of Dolpo are sheltered in addition by the high mountain country which extends eastwards from Kanjiroba Himal and therefore have an even drier climate than the southern parts.

Dolpo is a country of rolling hills, often smooth and round in outline but sometimes crowned with masses of red or yellow rock which in shape resemble the castle-like peaks of the Dolomites. During the monsoon months rain is only intermittent; more often the wide sunny sky is flecked with small white clouds, and as they bowl along in the western wind their restless blue shadows give a feeling of movement to the whole landscape. It is a different world from the steep ridges and deeply eroded valleys of the monsoon Himalaya.

Villages are few, and such as there are lie high, mostly between 13–14,000 ft. Cultivation is carried on up to 14,500 ft; the main crop is barley, but some wheat and buckwheat are also grown, and a very few potatoes and radishes. All the fields are irrigated. In many places the main tracks run for miles at easy gradients at altitudes over 16,000 ft, and in the summer there are numerous tented camps at these heights used by herdsmen grazing their yaks on the high alpine slopes. Although some of the mountain peaks are 21,000 ft in height they do not appear to be so high, for in the dry sunny climate their southern sides are free from snow.

Dolpo is almost but not quite treeless. A little *Pinus excelsa* and *Betula utilis* grow in the Sibu khola which runs down from Sya Gompa to Phijor, and also in the Barbung khola between Kakkot and Mukut. I have also seen a few small trees of *Betula* on inaccessible rocks near Simen. The great altitude at which most of this country lies would prevent the growth of trees even if the rainfall were sufficient, but I suspect that such few trees as may once have eked out a precarious existence at lower altitudes in favoured sites such as riversides have long ago been removed by man. On irrigated land around the villages there are a few small trees of *Salix*.

In northern parts of Dolpo two species of shrub dominate the treeless steppes; *Caragana brevifolia* and *Lonicera spinosa*. For mile after mile these are the only shrubs one sees in any numbers, though at about 16,000 ft they give way to a high altitude herb flora. In southern parts of the country, where conditions are rather less dry, other shrub species are more numerous: *Rhododendron antho-*

pogon, *R. lepidotum*, *R. nivale*, *Juniperus wallichiana*, *J. squamata*, *Rosa sericea*, *Potentilla fruticosa* and species of *Berberis*. Not much of Dolpo falls below 13,500 ft but where it does *Caragana brevifolia* is replaced by *C. gerardiana*, the altitude at which the change takes place being very constant.

There is some comparatively low country along the river in the Barbung khola between 11–12,000 ft. Here the soil is a sandy alluvium, and the vegetation consists largely of an *Artemisia*-grass steppe with some low bushes of *Caragana gerardiana* and *Cotoneaster* species. This is the only place in Dolpo where I have seen anything which can in any way be described as grass steppe, though species both of grasses and *Artemisia* are quite commonly found scattered among the *Caragana–Lonicera* shrubs. The grass species which I have collected in Dolpo are as follows: *Deyeuxia holciformis*, *Deyeuxia pulchella*, *Melica scaberrima*, *Melica jacquemontii*, *Poa poophagorum*, *Poa pagophila*, *Poa alpigena*, *Oryzopsis lateralis*, *Festuca ovina*, *Cymbopogon stracheyi*, *Danthonia cachemyriana*, *Orinus thoraldii*.

Between 13,500–15,000 ft there are often wide gravel flats along the streams. Here thickets of willow up to 10 ft high are common together with species of *Hippophae* and *Myricaria*. On the damp shady ground beneath them grow *Primula sikkimensis*, *Primula tibetica*, *Primula involucrata*, *Orchis stracheyi*, *Triglochin palustris*, *Gentianella paludosa*, *Pedicularis longiflora* var. *tubiformis*, and species of *Ranunculus*, *Epilobium*, and *Juncus*.

In some places along the rivers there are low cliffs formed of a soft sandy rock which erodes into fantastic flutings and pinnacles. On these rocks and sandy screes grow *Dicranostigma lactucoides*, *Incarvillea arguta*, *Silene moorcroftiana*, *Solmslaubachia fragrans*, *Morina polyphylla*, *Corydalis stricta*, and *Dracocephalum heterophyllum*.

Above 16,000 ft *Caragana brevifolia* and *Lonicera spinosa* begin to fade away, and the stony slopes are dotted with a discontinuous cover of high-altitude herbs. These herbs are almost all in some way adapted to resist the drought. Some form tight cushions which reduce transpiration (*Arenaria polytrichioides*, *Thylacospermum rupifragum*, *Androsace tapete*, *Androsace muscoidea*, *Potentilla biflora* var. *lahoulensis*, and several species of *Saxifraga in the Kabschia* group); or form tight-growing mats (*Allardia glabra*, *Calophaca nubigena*, *Primula minutissima*, *Saxifraga flagellaris*). Some are clothed in woolly hairs (*Eriophyton wallichianum*, *Allardia tomentosa*, *Tanacetum nubigenum*, *Saussurea sacra*, *Saussurea gnaphalodes*, and species of *Leontopodium*); or have leaves which though big lie flush with the ground (*Phlomis rotata*, *Oreosolen wattii*, *Incarvillea grandiflora*).

Many of the plants which grow on the screes at high altitudes do not adopt these protective devices for their leaves. Instead they grow very long roots in order to penetrate down to the moisture beneath the stones, and since the plants themselves are mostly quite small and their roots may be several feet long they appear very disproportionate. Species which adopt this method of survival include *Glechoma nivalis, Arenaria glanduligera, Cremanthodium nanum, Soroseris hookeriana, Scutellaria prostrata*, and several species of *Corydalis* and of the *Cruciferae* family.

The highest altitude in this country at which I have seen anything in the nature of a continuous sward of alpine vegetation is at 17,000 ft. Quite a number of scattered herbs manage to grow at 18,500 ft, the highest point I have reached here.

PART II

Forest Types

FOREST TYPES

It is obvious that in order to describe the tree and shrub vegetation of the whole of Nepal some degree of generalisation is necessary, despite the fact that the vegetation on any particular hillside often has characteristics which are difficult to fit into any generalised pattern. The difficulties are much increased where, as is often the case in Nepal, the original forest has been modified by man.

The following classifications are tentative. I hope that they may be of some help as an introduction to the forests of Nepal for those who do not know them. On the other hand I hope that those whose experience of these forests is greater, and who perhaps may disagree with my classifications, will in due course feel impelled to improve on them.

I have been much helped by the work of others. U. Schweinfurth's *Die horizontale und vertikale Verbreitung der Vegetation im Himalaya*[1] gives one a wide view of the various types of vegetation which occur throughout the whole of the Himalaya, and it is also indispensable as a work of reference for tracing authorities who have written about any particular part of the range.

The ecological notes in the introduction to A. E. Osmaston's *Forest Flora for Kumaon*[2] are admirably clear and succinct and can be applied almost without modification to the forests which occur in all the western parts of Nepal.

H. G. Champion's *Preliminary survey of the forest types of India and Burma*[3] has been of the greatest help to me. In my notes I have followed with certain modifications the classifications used in this survey. I understand from Sir Harry Champion that this original survey of 1936 has recently been revised by himself and others at Dehra Dun, and I should have liked to relate my own classifications to this revised version. At the date of writing, however, this revised version still awaits publication, and the references to Champion's types and page numbers, which I make at the head of the various forest types, all refer to his publication of 1936.

It would be an impertinence for an amateur botanist working alone to suppose that he could improve upon classifications made by trained foresters with all the resources of the Indian Forest Department behind them. Some explanation therefore is necessary as to why I have diverged in certain respects from Champion's forest types. By 1936 the Western Himalaya were very well known, and I have found that Champion's West Himalayan forest types fit well much of the forest which occurs in western parts of Nepal. Such variations in classification as I have made here are not of much significance. His East Himalayan forest types on the other hand do not fit so well the forests which occur in eastern parts of Nepal. In this respect one must point out that although the forests of Sikkim are well known they do not come under the control of the Indian Forest Department. K. C. R. Choudhury in 1951[4] produced an account of the Sikkim forests, but in this account he adhered strictly to the classifications used by Champion, and I think it is fair to say that definitive classifications of the East Himalayan montane forests have yet to be made and must await the further exploration of Bhutan and the Assam Himalaya. In writing of the forests of East Nepal, therefore, I have felt free to adapt Champion's classifications wherever I found it convenient to do so.

Those who confine the use of the word 'forest' to trees which form a continuous high canopy are warned that in these notes I used the word in a much wider sense to include all forms of woodland.

In listing the species which occur in the various forest types I have divided them into (1) trees which form the top canopy, (2) trees forming a second story, (3) smaller trees and shrubs, (4) climbers and epiphytes. In fact a number of species can often be found in more than one category, e.g. many epiphytes often also occur as small shrubs. Within the categories I have not followed an alphabetical order, but have attempted to group like with like, e.g. species of the same family, evergreens, deciduous species. I must confess that I have not been entirely systematic about this, because in some cases I have placed species next to each other for no better reason than that one commonly sees them so placed in the field.

FOREST TYPES

SAL FOREST

The sal forests of India yield valuable timber, and they have been intensively studied. Champion refers to the forests of Kumaon as moist sal forest and the forests of N. Bengal as wet sal forest. With the whole 500-mile length of Nepal intervening between the two areas one might expect a considerable difference in the type of forest found in them. Champion, however, states (at p. 86) 'The general appearance and composition of the main canopy (of the wet sal forest) is indistinguishable from the moist sal forest. *Shorea robusta* remains the gregarious dominant and the associated species are practically the same . . . the undergrowth is also very closely similar.'

The distinction between the two types of forest appears therefore to be a fine one, and such characteristics as can be used to distinguish them doubtless fade away progressively as they approach their meeting point in Nepal. I point out below such small differences as I have seen in sal forests in the eastern and western parts of Nepal, but I do not distinguish between Champion's two types.

I think it necessary, however, to distinguish in Nepal between the bhabar and terai sal which grows on flat ground and where trees attain a considerable height, and the very much smaller trees of the hill sal found in the foothills and the lower valleys of the Midlands.

Bhabar and terai sal forest

The sal tree, *Shorea robusta*, is widely distributed throughout northern India and in the subhimalayan tracts. It is an extremely gregarious species and is only rarely found as a component of any other type of forest. Although its importance as a timber tree makes sal a worthy subject of study for the forester, sal forest is not at all rich in associated species and for the botanist it is one of the least interesting forest types in Nepal.

At one time sal forest must have covered much of the flat terai lands which lie between the bhabar and the Indian border. Most of these lands have now been cleared for agriculture, and such few remaining patches of forest as I have seen there do not differ much in composition from the bhabar sal.

Large tracts of the bhabar country are waterless for many months of the year and therefore are uninhabited. Sal forest covers much of this country and of the dun valleys folded away within the outer foothills. If in the spring one flies over these forests it is easy to see the dominance of the sal, for at that time of year the sal is in flower and its sprays of white flowers and fresh green leaves are clearly visible from the air. Its dominance is broken only along the streams and rivers where the leafless trees of the deciduous riverain forest can be seen, or in dips and gulleys where the darker green foliage reveals the presence of tropical evergreen forest.

In April most of the watercourses are dry, and big expanses of gravel shimmer in the midday heat, but during the monsoon months these places are subject to flooding. The presence of other forest types along these watercourses is accounted for by the fact that sal will not grow either on waterlogged ground or on recent alluvium.

Compared to the great dark Dipterocarp forests of South-East Asia the sal forests of Nepal are light and open. Trees of 150 ft can occasionally be found, but in general the height of the forest is not much more than 80 ft. In many places the bigger trees have been felled, and the proportion of sal to other species may have been altered as well.

Epiphytes and climbers are scarce. In some places there is a dense green mass of shrubs and seedling trees beneath the trees, but in others the forest is carpeted only with dry brown leaves. To a certain extent the latter state may result from fire, for these forests are frequently burned in the spring, and it is common at that time of year to see a glowing line of fire creeping across the leaf carpet beneath the trees.

In bhabar sal forest around Dharan in east Nepal I have noted the following species:

1 Shorea robusta, Terminalia myriocarpa, Terminalia chebula, Terminalia belerica, Terminalia tomentosa, Schleichera trijuga, Dillenia pentagyna, Amoora decandra, Stereospermum suaveolens, Anogeissus latifolia, Adina cordifolia, Sterculia pallens, Lagerstroemia parviflora, Eugenia jambolana, Lannea grandis.

2 Careya arborea, Semecarpus anacardium, Ehretia laevis, Mallotus philippinensis, Glochidion velutinum, Croton oblongifolius, Litsea salicifolia.

3 Zizyphus rugosa, Antidesma diandrum, Clausena excavata, Leea sp.

The composition of bhabar sal forest north of Nepalganj in West Nepal is very similar:

1 Shorea robusta, Terminalia tomentosa, Terminalia belerica, Terminalia chebula, Adina cordifolia, Anogeissus latifolia, Lannea grandis, Schleichera trijuga, Eugenia jambolana.

2 Mallotus philippinensis, Semecarpus anacardium, Dillenia pentagyna, Kydia calycina, Aporosa dioica, Casearia tomentosa, Buchanania latifolia.

3 Ardisia humilis, Zizyphus rugosa, Clausena sp., Barleria cristata.

4 Spatholobus roxburghii, Bauhinia vahlii.

Hill sal forest

Sal does not grow usually much above 3,500 ft, though exceptionally it can be found at almost 5,000 ft. Hill sal forest grows on the outer foothills, and penetrates for a considerable distance up the main river valleys. Near Jajarkot in West Nepal sal grows in the Bheri valley more than 50 miles in a direct line from the plains.

Sal trees of the hill forest are only 40–50 ft tall. This diminution in size in comparison with terai and bhabar sal is probably due not to the increased altitude but to the shallowness of soil on most of the hillsides. In the hills north of Dhangarhi in West Nepal at over 3,000 ft there are some magnificent stands of sal 120 ft tall, but there the trees are growing on the deep soil of an isolated plateau.

Until recently the prevalence of malaria has discouraged the hill people from settling at the altitudes at which sal grows, and for this reason much hill sal remains uncut on the outer foothills and in the dun valleys. Such hill sal as penetrates into the densely populated Midland areas has suffered greatly from lopping and felling. In fact the sal is better able to withstand this harsh treatment than most of its associated species, many of which have disappeared from these hill forests leaving an almost pure sal community. Stunted pole-like sal trees forming much-lopped open forest are a common sight on the red laterite slopes of many Midland valleys.

In West Nepal the hill sal is usually superseded at about 3,500 ft by *Pinus roxburghii* forest. In a few places I have seen the pine pressing down to about 2,000 ft and forming a curious mixed forest with the sal. The pine here is scattered discontinuously amongst the predominant sal, but because its 80-ft trees easily overtop the 50-ft sal trees the forest when seen from a distance appears to consist predominantly of pine. This mixture seems to be anomalous, and the forest here is probably in a state of transition.

In West Nepal the sal is mostly displaced by subtropical deciduous hill forest on dry south faces. In the wetter parts of Central and East Nepal, however, the sal is largely confined to these dry south faces, for here the moister slopes are usually covered with some form of *Schima–Castanopsis* forest.

In the hills north of Janakpur in Central Nepal I have noted the following species in hill sal forest:

1 Shorea robusta, Lagerstroemia parviflora, Anogeissus latifolia, Adina cordifolia, Semecarpus anacardium, Bauhinia variegata, Dillenia pentagyna, Buchanania latifolia.

2 Nyctanthes arbortristis, Kydia calycina, Leucomeris spectabilis, Glochidion velutinum, Symplocos racemosa.

3 Hamiltonia suaveolens, Phoenix humilis, Indigofera pulchella, Flemingia strobilifera.

4 Bauhinia vahlii, Spatholobus roxburghii.

In the hills north of Nepalganj in West Nepal I have noticed very little difference in the composition of the hill sal forest, except that here *Terminalia tomentosa* and *Anogeissus latifolia* are a good deal more abundant than in the east.

TROPICAL DECIDUOUS RIVERAIN FOREST

Champion: Gangetic tropical moist deciduous riverain forest, p. 112.

This type of forest is found throughout the Gangetic plain, mostly in the sub-himalayan tracts of the United Provinces and Bihar.

In Nepal it is found frequently along streams of the bhabar and dun valleys. In these places there are often three distinct bands of forest along the streams; khair and sissoo at the water's edge and on gravel islands in the middle of the watercourses; deciduous riverain forest on riverside terraces with a more stabilised soil; behind this on rising ground the great blanket of sal forest stretching away into the distance.

In the spring the trees in this deciduous riverain forest are leafless except for a few such as *Mallotus philippinensis* and *Eugenia jambolana*. These leafless trees after five almost rainless months rather surprisingly burst into flower at the beginning of the hot weather. The forest composition is very mixed, but much the most prominent species is *Bombax malabaricum*, the simal tree. With its great buttressed trunk and scarlet flowers on leafless branches it is a fine sight in February. It is from the light buoyant wood of this tree that all the dugout canoes which are used as ferry boats in the big rivers of Nepal are made. Other species common here are *Holoptelea integrifolia*, *Schleichera trijuga*, *Ehretia laevis*, *Trewia nudiflora*, and *Garuga pinnata*. *Eugenia jambolana* often occurs close to the water.

Sometimes the forest forms a continuous canopy, but much more frequently the canopy is discontinuous. With water conveniently at hand these flat terraces on which the forest grows are much used for winter grazing, especially in western parts of the country. Cattle and sheep from the Jumla area descend during the winter into the low-altitude forests north of Nepalganj in much greater numbers than is customary elsewhere in Nepal. In consequence the deciduous riverain forest found in these places is usually much cut and burnt, and often is reduced to isolated trees forming a savannah-like forest on open grazing ground. Where the forest cover is discontinuous, scrub consisting of *Zizyphus jujuba*, *Clerodendrum infortunatum*, *Adhatoda vasica*, or *Colebrookea oppositifolia* is often found. At the present time, when due to pressure of population in the hills there is a considerable immigration of hillmen into the low country, these flat places close to the water are usually the first to be cleared for agriculture, and crops are often grown with the bigger trees ringbarked and still standing.

Because of the rather broken nature of the forest its composition is not easy to list, and a number of the species listed below are undoubtedly only secondary. Some also are not typical of the forest in its drier form, and are usually associated with water (*Albizzia lucida*, *Albizzia procera*, *Eugenia jambolana*). Between Janakpur and the western end of the country I have not noticed very much difference in the composition of this type of forest. In the extreme east, around Dharan and Bhadrapur, this type of forest seems to be much less common.

I have noted the following species:

1 Bombax malabaricum, Adina cordifolia, Schleichera trijuga, Holoptelea integrifolia, Lannea grandis, Ehretia laevis, Lagerstroemia parviflora, Sterculia villosa, Sapium insigne, Stereospermum suaveolens, Garuga pinnata, Careya arborea, Trewia nudiflora, Eugenia jambolana, Acacia catechu, Albizzia procera, Albizzia lucida.

2 Mallotus philippinensis, Croton oblongifolius, Holarrhena antidysenterica, Croton caudatus, Streblus asper, Cassia fistula, Aporosa dioica, Bridelia retusa, Alangium salviifolium.

3 Zizyphus jujuba, Colebrookea oppositifolia, Pogostemon plectranthoides, Adhatoda vasica, Maesa indica.

4 Caesalpinia digyna, Acacia caesia, Acacia megaladena, Acacia concinna, Hiptage madablota, Vallaris solanacea, Dalbergia volubilis, Merremia vitifolia, Natsiatum herpeticum, Tinospora cordifolia.

TROPICAL EVERGREEN FOREST

Champion: N. Bengal tropical evergreen forest, p. 51. Subhimalayan tropical semi-evergreen forest, p. 63. E. subhimalayan wet mixed forest, p. 119.

These forest types of Champion overlap and share a number of constituent species. They occur in Nepal, but because of their limited extent here it is not easy to distinguish between them. I think it more convenient to divide such forest in Nepal by altitude and geographical location as follows:

1 Tropical evergreen forest.
   Below 3,000 ft in the terai, bhabar, dun valleys and outer foothills of East, Central, and West Nepal.

2 Subtropical evergreen forest.
   Between 3–5,500 ft in areas of high rainfall on the outer foothills of East Nepal.

3 Subtropical semi-evergreen hill forest.
   Between 2–5,500 ft in areas of high rainfall at the base of the mountain ranges of East and Central Nepal.

In this section I deal only with tropical evergreen forest.

Extensive tropical evergreen forest such as is found in Assam and the Duars is not present in Nepal. J. M. Cowan, in writing of the forests of Kalimpong, which lie some 50 miles to the east of Nepal, states 'At the lower elevations, when the rainfall exceeds 180 in. (4,572 mm), evergreen species predominate, but where the rainfall is less than 160 in. (4,064 mm) most of the trees are deciduous, shedding their leaves in the hot season before the rains.'[5]

If this be so one must assume that in no part of the outer tropical zone of Nepal does the rainfall exceed 160 in., for throughout the whole country including the extreme east the forests of the outer tropical zone are predominantly deciduous or semi-deciduous. Tropical evergreen forest occurs only as narrow strips along water courses or in gulleys, and it has a marked preference for shady north-facing sites. Most of the surrounding country is covered with sal forest, and if one looks down into tropical forest from a hillside above one can often trace the course of a stream by following the dark green foliage as it twists and turns among the lighter green of the sal forest. These strips of evergreen forest are not usually more than 100 yds in breadth, and may be very much less. It should, however, be remembered that here, as in almost all parts of Nepal, the effects of fire must be allowed for. In Assam the growth of sal forest in some places is deliberately encouraged by judicious burning, for without such burning the forest there would be almost exclusively evergreen. It may be that in Nepal, or at any rate in its eastern parts, tropical evergreen forest is confined to damp and shady sites not so much by reduced rainfall as by the fires which sweep the country periodically.

East Nepal

I have not visited true terai forest here, but between the Kosi and the Mechi rivers I have seen tropical evergreen forest occurring in a few places in bhabar forest, and much more commonly in shady ravines and north-facing gulleys in the outer foothills.

This evergreen forest varies a good deal in appearance. In some places the evergreen trees attain a height of at least 150 ft which may bring their tops level with the surrounding sal forest despite the fact that their bases are deep in a gulley. In this tall dark forest there is usually a definite understory of smaller trees but few shrubs or climbers. In other places the general height of the trees does not exceed 60 ft, and beneath them there may be a dense growth of shrubs and climbers. Bamboos, palms and *Pandanus furcatus* are often abundant in such places, and thickets of thorny *Calamus* species occur in a few localities.

Of the larger trees *Michelia champaca* is perhaps the tallest and most prominent. Several species of the family *Lauraceae* are usually present, and so also are
species of *Eugenia*. Above 2,000 ft small stands of *Castanopsis tribuloides, Castanopsis indica,* and *Lithocarpus spicata* var. *brevipetiolata* sometimes occur, but they
are not typical of the forest. An odd tree of *Quercus glauca* can also sometimes be
found. Almost all the component tree species of the forest are East Himalayan
in distribution, and many of them occur in Nepal only in these wet shady places.

In the lower story *Murraya exotica, Micromelum integerrimum,* and various
species of the family *Araliaceae* are common, and in the spring the bright
flowers of *Phlogacanthus thyrsiflorus* and *Eranthemum nervosum* stand out amongst
the sombre evergreen foliage.

Sometimes the forest includes an element of tall deciduous trees, particularly
along the streams. Here are found species such as *Cedrela toona, Garuga pinnata,
Duabanga sonneratioides, Acrocarpus fraxinifolius,* and several *Albizzia* species.

In this type of forest east of the Kosi river I have noted the following species:

1 Cryptocarya amygdalina, Machilus villosa, Litsea salicifolia, Litsea oblonga,
Phoebe lanceolata, Actinodaphne angustifolia, Cinnamomum species, Michelia
champaca, Mangifera sylvatica, Eugenia jambolana, Eugenia frondosa, Dysoxy-
lum binectariferum, Dysoxylum procerum, Carallia lucida, Bassia butyracea,
Baccaurea sapida, Pterigospermum acerifolium, Horsfieldia (Myristica) kingii,
Knema (Myristica) linifolia, and species of strangling Ficus.

Quercus glauca, Castanopsis tribuloides, Castanopsis indica, Lithocarpus
spicata var. brevipetiolata.

Acrocarpus fraxinifolius, Garuga pinnata, Cedrela toona, Duabanga sonneratioides, Anthocephalus cadamba, Albizzia chinensis, Albizzia lebbek, Albizzia
lucida.

2 Actinodaphne obovata, Litsea polyantha, Eriobotrya elliptica, Turpinia pomi-
fera, Meliosma simplicifolia, Aphania rubra, Ehretia wallichiana, Ostodes
paniculata, Antidesma acuminatum, Cleidion javanicum, Sarauja roxburghii,
Casearia graveolens, Pandanus furcatus, Gynocardia odorata, and species of
Ficus.

3 Villebrunea integrifolia, Vernonia talaumifolia, Vernonia volkameriaefolia,
Ardisia humilis, Solanum crassipetalum, Phlogacanthus thyrsiflorus, Eranthemum nervosum, Pseuderanthemum indicum, Sterculia coccinea, Maesa
chisia, Trevesia palmata, Macropanax undulatum, Heteropanax fragrans,
Micromelum integerrimum, Murraya exotica, Morinda angustifolia, Garcinia

xanthochymus, Clerodendrum nutans, Cycas pectinata, and species of Mussaenda, Leea, Boehmeria, Calamus, and Wallichia.

4 Pothos cathcartii, Mucuna macrocarpa, Sabia paniculata, Trachelospermum lucidum, Melodorum bicolor, Poikilospermum lanceolatum, Jasminum pubescens, Polypodium coronans, and species of Raphidophora, Vitis, and epiphytic orchids.

Central Nepal

I have seen in the dun valleys and outer foothills of the country which lies between Janakpur and the Nayarani river tropical evergreen forest very similar in general appearance to that which I have described for East Nepal. It is found in similar positions in gulleys and along streams in the sal forest, but its composition appears to be a good deal less rich in species.

My experience of the area is insufficient to justify recording anything more than a suspicion that a number of the species listed above as occurring in tropical evergreen forest in East Nepal do not extend their range west of the Kosi river.

West Nepal

Before dealing with West Nepal it may be helpful to consider the occurrence of tropical evergreen forest in Kumaon. Champion makes no mention of such forest here or anywhere else in the Western Himalaya, but Osmaston refers to evergreen forests in swamps and moist localities along the banks of the terai streams and to a more limited extent in the outer ranges up to 2,500 ft.[6]

I have not visited true terai forest in West Nepal, but I have seen tropical evergreen forest growing here on the outer ranges. Its composition is very similar to that described by Osmaston for Kumaon forest. The two parallel Siwalik ridges which lie north of Nepalganj are covered at altitudes between 1–3,000 ft on their southern slopes with subtropical deciduous hill forest and on their northern slopes with sal forest. Amongst this sal forest there are numerous strips of evergreen forest in places where gulleys and watercourses occur.

The evergreen tree most constantly found in these places is *Eugenia jambolana*, but *Phoebe lanceolata*, *Mangifera sylvatica*, and a species of *Diospyros* are also quite common. Where the trees are big and the canopy continuous, climbers and undershrubs are few, but in lighter places *Murraya exotica*, *Ardisia humilis*, and *Eranthemum nervosum* are common undershrubs and *Trachelospermum lucidum* and *Jasminum pubescens* common climbers. In wetter places *Cedrela toona* and species of *Albizzia* occur, and in one locality near the Bheri river I have seen a number of trees of *Duabanga sonneratioides*.

In tropical evergreen forest in West Nepal I have noted the following species:

1. Eugenia jambolana, Mangifera sylvatica, Phoebe lanceolata, Machilus species, Bassia butyracea, Xylosma longifolium, Ormosia glauca, Acer oblongum, Cedrela toona, Duabanga sonneratioides, and species of Albizzia, Ficus, and Diospyros.

2. Olea glandulifera, Viburnum punctatum, Stranvaesia nussia, Mallotus philippinensis, Bischofia javanica, Macaranga denticulata, Alstonia scholaris, Murraya exotica, Heynea trijuga, Sageretia oppositifolia.

3. Ardisia humilis, Phlogacanthus thysiflorus, Eranthemum nervosum, Petalidium barlerioides, and species of bamboo.

4. Holmskioldia sanguinea, Jasminum pubescens, Sabia paniculata, Trachelospermum lucidum, and species of Vitis and Raphidophora.

It can be seen that this West Nepal forest is poor in species and contains quite a high proportion of deciduous trees, and I hesitate to place it under the same forest type as the very much richer tropical evergreen forest of East Nepal.

SUBTROPICAL EVERGREEN FOREST

Champion: See my note at the head of tropical evergreen forest. This forest type has affinities with the *Ostodes* association included by Champion as a subtype of Bengal subtropical hill forest, p. 200.

I have seen this type of forest only between 3–5,500 ft in localities of high rainfall on the outer foothills between the Kosi and the Mechi rivers in the extreme east of Nepal.

Ostodes paniculata is often abundant in this forest. The species can attain a height of 50 ft, but frequently it survives only in a much lopped form. *Eugenia tetragona* also is common here, and at about 5,000 ft near the Mechi khola the crest and northern slope of the outermost ridge are covered for some miles with forest in which these two species predominate to the almost complete exclusion of any others. The trees of *Eugenia* are moss-covered and about 50 ft tall, forming a close canopy. The trees of *Ostodes* grow beneath them to a height of about 25 ft. I have never seen forest of such a kind elsewhere in Nepal, and perhaps the forest composition here has been altered by selective felling in the past, though there are now no signs of this having taken place. The following list of species I noted growing in the forest in this locality show that it is unusually poor in species:

1 Eugenia tetragona, Acer thomsonii, Machilus species, Castanopsis indica.

2 Ostodes paniculata, Leucosceptrum canum, Eurya acuminata.

3 Maesa chisia, Ardisia macrocarpa.

4 Hymenopogon parasiticus, Vaccinium vaccinaceum, and species of Hoya, Raphidophora, and epiphytic orchids.

In other places this type of forest is far more mixed, and although the *Eugenia* and *Ostodes* remain common so also are species such as *Talauma hodgsonii, Drimycarpus racemosus, Lithocarpus spicata* var. *brevipetiolata*, and a number of species of the family of *Lauraceae*. The forest is very predominantly evergreen, but a few tall deciduous trees may be present such as *Cedrela toona* and species of *Albizzia*. The list of species I have noted in this forest is as follows:

1 Eugenia tetragona, Eugenia ramosissima, Drimycarpus racemosus, Dysoxylum procerum, Acer oblongum, Acer thomsonii, Phoebe lanceolata, Cryptocarya amygdalina, Machilus villosa, Cinnamomum species, Turpinia nepalensis, Bassia butyracea, Lithocarpus spicata var. brevipetiolata, Helicia erratica, Macaranga pustulata.

Alnus nepalensis, Erythrina suberosa, Cedrela toona, Albizzia lebbek, Albizzia chinensis.

Schima wallichii, Castanopsis tribuloides, Castanopsis indica.

2 Ostodes paniculata, Talauma hodgsonii, Symplocos spicata, Eurya acuminata, Laportea sinuata, Leucosceptrum canum, Miliusa macrocarpa, Mahonia napaulensis, Casearia graveolens, Amoora decandra, and species of Sarauja.

3 Ardisia macrocarpa, Clerodendrum colebrookeanum, Solanum crassipetalum.

4 Vaccinium vaccinaceum, Agapetes sikkimensis, Hymenopogon parasiticus, and species of Raphidophora, Aeschynanthus, Hoya and epiphytic orchids.

Despite the presence of a few trees of *Schima* and *Castanopsis* this forest is very different in composition from *Schima–Castanopsis* forest.

TERMINALIA FOREST

Champion: North Indian moist Terminalia forest, p. 93.

Species of *Terminalia* in Nepal are common as components of tropical and subtropical forest, but they predominate in these forests only rarely and in areas of very limited extent.

The following species have been recorded from Nepal: *Terminalia tomentosa, T. chebula, T. belerica, T. myriocarpa.*

Terminalia myriocarpa is common in the bhabar and dun valleys of the eastern part of the country, and it is one of the few big trees of the forest here which comes into flower in the middle of the monsoon. Usually it is only a component of forest in which sal predominates, but there are limited areas where small patches of what can loosely be called *Terminalia* forest occur. For example, there are places in the Rapti valley in Central Nepal where the sal gives way to forest in which *Terminalia myriocarpa* predominates, with *Terminalia tomentosa* and *Terminalia belerica* also present. The trees here are 100 ft tall, and are accompanied by some *Eugenia jambolana, Lagerstroemia parviflora, Dillenia pentagyna, Adina cordifolia* and *Cedrela toona.*

Terminalia chebula, T. belerica and *T. tomentosa* are widespread in tropical and subtropical forest in western as well as eastern areas of Nepal, and can also be found in the lower river valleys of the Midlands. *T. tomentosa* in particular is very common in the subtropical deciduous hill forest.

In a place a few miles west of the junction of the Mayangdi khola with the Kali Gandaki in Central Nepal at an altitude of 3,000 ft forest consisting almost entirely of *Terminalia tomentosa* covers some acres of ground. The trees are about 50 ft tall, and when seen in mid-May they have a very curious appearance. All the other trees in the vicinity are either evergreen or else are by now in full green leaf. Only the *Terminalia* trees stand leafless and apparently lifeless beneath the hot May sun, like a forest stricken by a defoliant spray.

I have seen similar pure *Terminalia tomentosa* forest in the Bheri valley in West Nepal, and in both places there was very little regeneration beneath the trees. Perhaps these forests as they now stand are only the remnants of a much more mixed type of forest from which the other species have been culled.

DALBERGIA SISSOO–ACACIA CATECHU FOREST

Champion: Khair–Sissoo forest, p. 181.

Both khair (*Acacia catechu*) and sissoo (*Dalbergia sissoo*) are widespread along the rivers of Northern India.

Forest composed of khair and sissoo is common on new alluvium along the streams in the bhabar and dun valleys of Nepal. The extent of the forest is limited by the extent of riverside gravel. The khair and sissoo consolidate the soil, and if a flood does not sweep this soil away again they may be superseded after some years by deciduous riverain forest.

In some places both species occur together, but frequently the forest is composed almost entirely of one species or the other. Trees of both species attain a height of about 50 ft, and often there is dense undergrowth beneath them of *Pogostemon plectranthoides* or *Colebrookea oppositifolia*.

The sissoo is limited almost exclusively to riverain sites. In March its new leaves are a vivid green at a time when most of the other riverain trees are leafless. Khair on the other hand is found as a component both of deciduous riverain forest and of subtropical hill forest, and in the latter type of forest it may grow on hot dry hillsides far from any water.

Khair and sissoo penetrate for a considerable distance into the hills. For example, they ascend the Karnali valley beyond the 4,000 ft contour almost to the confluence of the Humla and Mugu branches of the river. This locality is about 80 miles in a direct line from the plains.

SUBTROPICAL DECIDUOUS HILL FOREST

Champion: Northern tropical dry mixed deciduous forest, p. 147.

This type of forest is North Indian in distribution. In Nepal it occurs on the outer foothills at altitudes up to 4,000 ft, and also penetrates into the Midland areas up the big river valleys.

The forest is much more abundant in western than in eastern parts of Nepal.

In the country around Nepalganj and Dhangarhi in West Nepal the southern slopes of the outer foothills are covered almost exclusively with extensive forest of this type, and here sal forest is mostly confined to the northern slopes. In contrast the southern slopes of the outer foothills north of Dharan and Bhadrapur in East Nepal are covered with hill sal forest or with wetter forest in which *Schima wallichii, Castanopsis indica,* or *Ostodes paniculata* may be present. The deciduous hill forest is confined here almost entirely to dry south faces in the bigger river valleys.

The most prominent feature of subtropical deciduous hill forest is its state of leaflessness in the spring, and in this respect it ressembles tropical deciduous riverain forest. These two forest types in fact share a number of component species, and in some places one can find forest of a kind that is transitional between the two. In their most typical form, however, these forest types are quite distinct. The riverain forest grows on river terraces in the bhabar and dun valleys and contains trees of good height, and in particular much tall *Bombax malabaricum.* The hill forest grows on hot dry slopes where the soil is thin and the trees are only 40–50 ft tall. On hill slopes where the soil is richer and deeper deciduous hill forest is likely to be replaced by sal forest.

If in March when the trees are leafless one looks from a distance at a slope covered with deciduous hill forest much open ground is visible beneath the rather small trees, but when the trees are in leaf the canopy may be more or less continuous. The composition of the forest is very mixed except in places where *Anogeissus latifolia* predominates. On hot rocky slopes this species tends to form almost pure forest in some areas. In late February the bright red old leaves of this species are prominent, but by the end of March it is leafless.

Other trees typical of this forest are *Ehretia laevis, Terminalia tomentosa, Flacourtia indica,* and *Lannea grandis. Mallotus philippinensis* is one of the few evergreen species commonly present. The leafless flowering trees of *Bauhinia variegata, Ougeinia dalbergioides,* and *Alangium salviifolium* are colourful in March.

In the spring the cloudless skies of the outer foothills are filled with the haze of smoke from forest fires, for the graziers here regularly burn the forest to improve the grazing. In consequence there is usually little undergrowth and much grass beneath the trees of subtropical deciduous hill forest, and in many places the forest is very discontinuous and rather secondary in character. On the more open slopes shrubs such as *Woodfordia fruticosa* and *Rhus parviflora* are common.

71

In Central Nepal on the outer foothills which lie between Narayangarh and Janakpur I have noted the following species in this type of forest:

1 Anogeissus latifolia, Lagerstroemia parviflora, Adina cordifolia, Bauhinia variegata, Dalbergia latifolia, Acacia catechu, Ougeinia dalbergioides, Croton oblongifolius, Glochidion velutinum, Flacourtia indica, Ehretia laevis, Garuga pinnata, Aegle marmelos, Phyllanthus emblica, Mallotus philippinensis, Lannea grandis, Engelhardtia spicata, Terminalia tomentosa, Sapium insigne, Litsea sebifera, Buchanania latifolia, Kydia calycina, species of Wendlandia.

2 Woodfordia fruticosa, Alangium salviifolium, Rhus parviflora, Butea minor, Phoenix humilis.

3 Mimosa rubicaulis, Acacia pennata, Acacia megaladena.

There does not appear to be very much difference in the composition of this type of forest in western parts of the country. In the hills north of Nepalganj I noted the following species:

1 Anogeissus latifolia, Lagerstroemia parviflora, Terminalia tomentosa, Buchanania latifolia, Acacia catechu, Semecarpus anacardium, Bauhinia variegata, Engelhardtia spicata, Ougeinia dalbergioides, Flacourtia indica, Wendlandia species, Phyllanthus emblica, Sterculia villosa, Ehretia laevis, Aegle marmelos, Piptadenia oudhensis.

2 Hamiltonia suaveolens, Woodfordia fruticosa, Phoenix humilis.

SCHIMA-CASTANOPSIS FOREST

Champion: Bengal subtropical hill forest, p. 190.

The three *Castanopsis* species recorded from Nepal are all East Himalayan in distribution. *Castanopsis tribuloides* is the only one of them known to extend its range into the West Midlands of Nepal and into Kumaon. *Castanopsis indica* is fairly common in the East and Central Midlands but has not been recorded west of the Kali Gandaki. *Castanopsis hystrix* occurs in a few places in the East Midlands and has been recorded as far west as Okhaldunga. *Schima wallichii*, another East Himalayan species, is common in the East and Central Midlands. I have seen it as far west as the Mayangdi khola to the south of Dhaulagiri.

Schima–Castanopsis forest occurs between 2–6,000 ft. It is quite wrong to think of this forest as forming one continuous belt between these altitudes, for in Nepal the zone of maximum cultivation lies at this height. With numerous villages and terraced hillsides all that remains of the forest in many places are the coppiced stumps of *Castanopsis*, or a few *Schima* trees scattered with some secondary shrubs across an otherwise deforested hillside.

Sufficient forest survives for one to be able to see that at these altitudes *Schima–Castanopsis* forest must once have covered most of the north-facing and south-facing slopes in the wet country of the upper Arun and Tamur, and also much of the lower country to the south of Annapurna Himal. Elsewhere in the East and Central Midlands this type of forest is more restricted in its habitats and is virtually confined to north or west slopes, the south or east ones in the drier conditions prevailing here being covered with sal forest, *Quercus incana–Quercus lanuginosa* forest, or *Pinus roxburghii* forest.

Because of its damaged state and because it varies very much in composition at different altitudes it is not easy to give a generalised account of this forest. *Schima wallichii* and *Castanopsis indica* tend to predominate between 2–4,500 ft, *Schima wallichii* and *Castanopsis tribuloides* between 4,500–6,000 ft; but at all altitudes there may be numerous other species present and neither *Schima* nor *Castanopsis* may be particularly predominant. *Schima wallichii* and *Castanopsis indica* do not normally ascend above 6,000 ft, but *Castanopsis tribuloides* goes at least 1,000 ft higher, and between 6–7,000 ft on the fringe of the *Quercus lamellosa* belt *Castanopsis tribuloides* and *Castanopsis hystrix* occasionally can be found forming almost pure forest. This latter forest I refer to as a separate forest type.

Schima wallichii–Castanopsis indica forest

In the Pokhara district forest consisting predominantly of *Schima wallichii* and *Castanopsis indica* is extensive between 2,500–5,000 ft, on both south-facing and north-facing slopes.

In this very wet area sal forest is not extensive, being limited to a few south faces. Treeferns, *Pandanus*, and other species of the subtropical wet mixed forest are common in wet gulleys in the *Schima–Castanopsis forest*, and at about 5,000 ft this forest passes into a *Michelia–Laurel–Lithocarpus* mixture typical of the lower temperate mixed broadleaved forest.

Schima wallichii and *Castanopsis indica* here both attain a height of about 80 ft. The white flowering spikes of the *Castanopsis* are very prominent in the clear rainwashed sunshine of post-monsoon October. These two species are very dominant in the forest, but in its lower part there are also a number of tropical species present, such as *Mallotus philippinensis, Bombax malabaricum, Terminalia chebula* and *Eugenia jambolana*. In wet gulleys there are little pockets of moisture-loving species typical of the tropical evergreen forest which occurs in the outer foothills, such as *Michelia champaca, Anthocephalus cadamba, Dysoxylum procerum*, and *Heteropanax fragrans*. In this lower part of the forest *Castanopsis tribuloides* is comparatively rare.

Higher up the composition changes. *Schima wallichii* and *Castanopsis indica* remain dominant up to 5,000 ft but the associated species here are rather different: *Macaranga pustulata, Rhus succedanea, Ilex doniana, Engelhardtia spicata*; and the wet gulleys hold treeferns and *Pandanus furcatus*.

In this type of forest around Pokhara I have noted the following species:

> Castanopsis indica, Castanopsis tribuloides, Schima wallichii, Engelhardtia spicata, Bombax malabaricum, Shorea robusta, Myrica esculenta, Litsea oblonga, Macaranga pustulata, Rhus succedanea, Sapium insigne, Mallotus philippinensis, Heynea trijuga, Ilex doniana, Casearia graveolens, Terminalia chebula, Eugenia jambolana, Eugenia frondosa, Lithocarpus spicata, Quercus glauca.

Where this type of forest remains uncut, shrubs and climbers are few. In many places, however, it has been much lopped. To the south of Himal Chuli there is some almost pure *Castanopsis indica* forest at 5,000 ft, in which the lopped trees are only about 40 ft tall. In this rather open forest shrubs are more numerous, and include *Camellia kissi, Viburnum erubescens, Viburnum coriaceum, Eurya acuminata, Myrsine semiserrata, Osbeckia stellata, Pyrus pashia, Rhododendron arboreum*, and species of *Phyllanthus* and *Sarcococca*.

Schima wallichii–Castanopsis tribuloides forest

Schima wallichii–Castanopsis tribuloides forest in which *Castanopsis indica* is comparatively scarce is widespread in the Arun and Tamur valleys, both on north and south faces. It occurs between 2–6,000 ft, but most commonly between 4–5,500 ft. This type of forest in various stages of degradation also occurs in other parts of the East and Central Midlands, but in these areas it is usually confined to north or west faces.

The distinction between this forest subtype and the one previously described is not a rigid one. On the whole it can be said that above 4,000 ft *Castanopsis tribuloides* becomes more numerous and *Castanopsis indica* less so. *Schima wallichii* remains very numerous throughout, and in some places occurs without any *Castanopsis* in association with *Engelhardtia spicata, Rhus succedanea*, and other subtropical species. Perhaps this latter form of forest should be described as a separate subtype but in fact the forest at these altitudes has been so badly damaged that fine distinctions between its various forms seem to be unjustified. Even at its best the forest rarely forms a continuous canopy, and the presence of species such as *Callicarpa arborea*, *Macaranga pustulata*, and *Macaranga denticulata* reveal that it is often secondary in character.

In forest between 3–6,000 ft in the Tamur valley I noted the following species:

1 Castanopsis tribuloides, Castanopsis indica, Schima wallichii, Engelhardtia spicata, Alnus nepalensis, Lithocarpus spicata, Quercus glauca, Carpinus viminea, Eugenia frondosa.

2 Callicarpa arborea, Wightia speciosissima, Macaranga denticulata, Macaranga pustulata, Helicia erratica, Rhododendron arboreum, Lyonia ovalifolia, Rhus semialata, Rhus succedanea, Wendlandia species.

3 Maesa chisia, Eurya acuminata, Oxyspora paniculata, Pentapanax leschenaultii, Viburnum coriaceum, Cornus oblonga, Dobinea vulgaris, Atylosia mollis.

4 Vaccinium vaccinaceum.

Near Halesi on the Kosi in East Nepal there is some forest on a north face between 5–6,000 ft which indicates that the forest described immediately above probably would have been very different in its natural state. Here the slope is steep and much of it is strewn with large boulders, so that it is of little value for grazing or agriculture and the forest appears to have been left comparatively uncut. The trees are not more than 50 ft tall, and consist almost entirely of *Castanopsis tribuloides*. When I visited this place in October the village girls were busy collecting the *Castanopsis* nuts to press for oil. I saw no trees of *Castanopsis indica* or *Castanopsis hystrix*, and *Schima wallichii* though present was not numerous. Most of the secondary species one so often finds in *Schima–Castanopsis* forest were absent.

I noted the composition of this forest over about two miles of hillside as follows:

1 Castanopsis tribuloides.
   A few trees of Schima wallichiii, Litsea oblonga, Eugenia frondosa, Homalium nepalense, Juglans regia, Rhus succedanea. In wet gulleys Alnus nepalensis, Macaranga denticulata.

2 Rhododendron arboreum, Lyonia ovalifolia, Eurya acuminata, Camellia kissi.

3 Luculia gratissima, Oxyspora paniculata, and species of Boehmeria and Sarcococca.

SUBTROPICAL SEMI-EVERGREEN HILL FOREST

Champion: See note at head of tropical evergreen forest.

Many of the constituent species of this forest type occur also either in tropical evergreen forest or in subtropical evergreen forest as described above. I have no doubt that to a certain extent all three types overlap each other, and further study of the tropical forests of Nepal may show that these forests should be divided in some other manner. For the moment, however, I think it convenient to refer separately to this wet subtropical forest which can be found deep into the big mountains.

Subtropical semi-evergreen hill forest occurs between 2–5,500 ft, and differs in its geographical location from the two other forest types previously mentioned, for whereas they are confined to the outer foothills and the flat ground that lies beneath them this type is found at the base of the big mountains, chiefly in side valleys of the upper Arun and Tamur and in the vicinity of Pokhara. It is found mostly on north or west faces or in steep shady valley bottoms, and is not extensive.

Much of the surrounding country in these places is or has at one time been covered with some form of *Schima–Castanopsis* forest, and trees of *Schima wallichii*, *Castanopsis indica*, and *Castanopsis tribuloides* occur quite commonly in the subtropical semi-evergreen forest, but they are not dominant in it. Often at these altitudes the surrounding country has been cleared for cultivation and the forest only exists as narrow strips along the watercourses and in steep places. Secondary species are numerous, such as *Mallotus nepalensis*, *Macaranga pustulata*, and *Callicarpa arborea*, and I have no doubt that in the list below are included a number of other secondary species which would not be found in this forest if it were in an undisturbed state.

Two species very typical of this forest are *Pandanus furcatus* and *Cyathea spinulosa*, the latter species appearing to be much more numerous at the base of the big mountains than it is on the outer foothills. The canopy is usually very much broken, and tall deciduous trees often stand singly above the smaller evergreens; *Dalbergia hircina*, *Albizzia mollis*, *Cedrela toona*, *Erythrina suberosa*, and others.

These small pockets of wet forest provide suitable habitats for a number of

East Himalayan species which in Nepal are nearing the end of their westward range. For instance, *Podocarpus neriifolius, Cyathea spinulosa, Talauma hodgsonii,* and *Dysoxylum procerum* occur as far west as Pokhara in this type of forest.

In the Central and East Midlands I have noted the following species in subtropical semi-evergreen hill forest:

1 Cedrela toona, Albizzia chinensis, Albizzia lucida, Dalbergia hircina, Erythrina suberosa, Duabanga sonneratioides, Engelhardtia spicata, Ehretia wallichiana, Sapium baccatum, Dysoxylum procerum, Garuga pinnata, Gmelina arborea.
 Mangifera sylvatica, Bassia butyracea, Phoebe lanceolata, Phoebe attenuata, Beilschmiedia roxburghiana, Cryptocarya amygdalina, Cinnamomum species, Eugenia jambolana, Eugenia frondosa.
2 Ostodes paniculata, Macaranga pustulata, Macaranga denticulata, Mallotus nepalensis, Bischofia javanica, Cyathea spinulosa, Pandanus furcatus, Podocarpus neriifolius, Turpinia nepalensis, Spondias pinnata, Heynea trijuga, Symplocos spicata, Myrsine capitellata, Talauma hodgsonii, Sarauja napaulensis, Debregeasia wallichiana, Trema orientalis, Premna barbata, Picrasma javanica, Casearia glomerata, Casearia graveolens, Heteropanax fragrans, Alstonia scholaris.
3 Trevesia palmata, Leucomeris spectabilis, Maytenus rufa, Capparis multiflora, Micromelum integerrimum, Glochidion lanceolarium, Alstonia neriifolia, Phlogacanthus thyrsiflorus, Morinda angustifolia, Clerodendrum colebrookeanum, Clerodendrum nutans, Clerodendrum bracteatum, Randia fascisculata, Ardisia floribunda, Pavetta indica, Callicarpa macrophylla, Sterculia coccinea.
4 Schefflera venulosa, Ecdysanthera micrantha, Entada scandens, Trachelospermum lanceolatum, Dregea volubilis, Uncaria pilosa, Actinidia callosa, Pothos cathcartii, and species of Raphidophora, Vitis, and Aeschynanthus.

In West Nepal I have not seen any forest truly comparable to that described above, but certain deciduous tree species such as *Dalbergia hircina* and *Albizzia mollis* are quite common here at subtropical altitudes, particularly in association with water, and occasionally a few evergreen trees are also found in these places. In one locality on the Seti khola, west of the Karnali at 4,500 ft, I noted the following species:

Albizzia mollis, Ehretia acuminata, Dalbergia hircina, Stranvaesia nussia, Olea glandulifera, Viburnum punctatum, Cornus macrophylla, Rhus succedanea, Cocculus laurifolius, Trachelospermum lucidum, Actinidia callosa, Vitis species. Epiphytic ferns and orchids numerous.

77

PINUS ROXBURGHII FOREST

Champion: Himalayan subtropical pine forest, p. 205.

Pinus roxburghii, the longleaved pine, ranges from Afghanistan and Kashmir eastwards along the Himalayan chain to Bhutan. In Nepal its normal altitude range is between 3–6,500 ft, but in the outer foothills I have seen a few trees as low as 1,500 ft. In West Nepal in the dry upper valleys of the Karnali near Simikot and of the Bheri near Dunaihi the pine is quite abundant at 9,000 ft, but to find it at this height is unusual.

The West Midlands

Here between 3–6,500 ft there are extensive forests of pine both on south-facing and north-facing slopes. West of the Karnali river and south of Silgarhi the population at these altitudes is less dense, and in consequence the forest less cut, than in any other part of the country I have visited.

In its upper part the pine forest usually merges with oak forest, and in damp gulleys the oak may descend into the pine zone as low as 4,000 ft. In its lower part the pine often merges with sal or subtropical deciduous hill forest.

The top canopy of typical *Pinus roxburghii* forest is composed exclusively of pine. Trees are up to 100 ft tall, with straight unbranched trunks for much of this height. There is no second story of smaller trees, no climbers, and very few undershrubs. The trees are widely spaced and much of the ground beneath is clothed only with a carpet of brown needles. This lack of associated species is in large part due to the frequent fires here. Almost all the trees bear at their base the marks of past fires.

In damp gulleys in the pine forest one may find a few trees of *Quercus incana* or *Quercus lanuginosa* with some of their associated species such as *Rhododendron arboreum* or *Lyonia ovalifolia*, but these trees are not typical of the pine forest and are really only intrusions from another type of forest. Similarly, subtropical trees such as *Engelhardtia spicata* and *Terminalia* species sometimes spread up into the lower part of the pine forest from the deciduous hill forest. In typical pine forest in the West Midlands I have noted only the following species:

1 Pinus roxburghii.

2 Inula cappa, Woodfordia fruticosa, and species of Phyllanthus, Indigofera, and Wendlandia.

In places near the junction of the Bheri and the Karnali rivers at an altitude of about 2,000 ft there is an altogether different kind of forest association in which *Pinus roxburghii* has a prominent but not a dominant position. Seen from a distance the forest here appears to be composed very largely of pine, but a closer inspection reveals that the pine trees are widely spaced at intervals of 50 yds or so, and that the intervening forest has a composition typical of hill sal forest. The pine overtops the sal by 30–40 ft. It seems that the forest here is in a transitional state.

The Central Midlands

In this part of the country *Pinus roxburghii* is largely confined to situations where conditions are drier than average. The big river valleys provide many suitable habitats. Here the vegetation has been much altered by the activities of man, and these valley pine forests are much burnt, and frequently adjoin deforested grass slopes. Above the pine is often oak or *Pinus excelsa*. At its lower end the pine often runs into a dry valley bottom where only *Euphorbia royleana* and a few windblown shrubs manage to survive.

In wetter conditions elsewhere in the Central Midlands the pine appears unable to compete with broadleaved species. It is, however, not uncommon on south-facing slopes, or on ground which at one time has been under cultivation. In the very wet hills north and east of Pokhara it is absent even from the south faces.

The East Midlands

Pinus roxburghii is scarce here. It occurs in limited amounts in the lower parts of the Arun and Tamur valleys, but it is almost completely absent even from south faces in the wetter upper parts.

QUERCUS INCANA–QUERCUS LANUGINOSA FOREST

Champion: Ban oak forest (*Quercus incana*), p. 231. From Champion's notes it is evident that *Quercus lanuginosa* though present in Garhwal and Kumaon is not an important element in the forest there.

Quercus incana Roxb. has also been called *Quercus leucotrichophora* A. Camus; *Quercus lanuginosa* D. Don has also been called *Quercus lanata* Smith. In these notes

I refer to the two oaks as *Quercus incana* and *Quercus lanuginosa* respectively in the belief that whether correct or not these names are the ones most widely used at the present time.

The two species are generally considered to be distinct. G. King, for example, writes of *Quercus lanuginosa* as follows: 'Nobody who has ever seen the two growing would think of uniting this with *Quercus incana*, and even leaf specimens of this can be distinguished in the herbarium by their plentiful rufous tomentum, the vestiture of *incana* being more scanty and of a pale grey colour.'[7]

Hiroo Kanai, however, writes of the gatherings made by Japanese collectors in East Nepal, Sikkim, and Bhutan as follows: '*Quercus incana* and *Quercus lanuginosa* are hardly distinguishable, though they are separated from each other by the colour of leaf.'[8]

In West Nepal where the two species can often be found growing together I have myself found them to be readily distinguishable on the grounds given by King and by the *Flora of British India*. In the field the characteristics which I have found most useful to separate them are the densely tawny woolly young branchlets of *Quercus lanuginosa* and the silvery grey tomentose young branchlets of *Quercus incana*.

Quercus lanuginosa is recorded from Garhwal eastwards through Nepal and Bhutan to China. *Quercus incana* extends from Swat through the NW. Himalaya to Nepal. There appear to be no records in the herbaria at Kew or at the British Museum of the occurrence of *Quercus incana* in Sikkim or Bhutan, but there are two gatherings from the Burma–Assam border made by Kingdon Ward which have been named as this species, and the *Flora of British India* states that it occurs elsewhere in Burma. N. L. Bor, in his account of the vegetation of the Assam Himalaya north of Tezpur,[9] states that *Quercus incana* occurs in this area, but this species is not included in the list made by K. Biswas of the plants collected by Bor from here.[10]

I have spent some time in the field making observations on these two oaks in Nepal. West of the Karnali *Quercus incana* is much the more abundant of the two, as it is in Kumaon; it grows at suitable altitudes on all faces. *Quercus lanuginosa* is here confined to limited areas, almost always on south faces. In the Central Midlands, however, *Quercus lanuginosa* is undoubtedly the more widespread of the two species; for example, in the vicinity of the Nepal valley *Quercus lanuginosa* is very common, but I cannot remember ever having seen *Quercus incana* here at all. In the East Midlands all the oak at the relevant altitudes that I have inspected has consisted of *Quercus lanuginosa*, and although

I have for some years been on the lookout for *Quercus incana* here I have failed to find it.

My own view is that *Quercus incana* ceases to be of much importance as a component of the medium altitude oak forests of Nepal once one has moved eastwards across the Kali Gandaki, but in view of the difficulty expressed by Hiroo Kanai in distinguishing gatherings made in the Eastern Himalaya, and of the uncertainty of the range of *Quercus incana* eastward, I refer in these notes to forest composed of either or both of these oaks as *Quercus incana–Quercus lanuginosa* forest without making any further distinction between the two species.

Quercus incana–Quercus lanuginosa forest is extensive in the West Midlands where together with *Pinus roxburghii* it covers most of the medium-altitude hillsides which are not under cultivation. The oak forest predominates between 5,500–8,000 ft, but in the gulleys it presses down into the pine forest to 4,000 ft. In some places the lower part of the oak forest mingles with the subtropical deciduous forest, and occasionally I have seen it pass directly into sal forest at about 4,000 ft. Normally, however, there is a belt of pine intervening between the oak and the sal.

In the Central Midlands the oak is less widespread, and is confined mostly to south-facing slopes or the sides of the big river valleys where the conditions are drier than in the surrounding country. In these valleys it often associates with *Pinus roxburghii* below and *Pinus excelsa* above. Where it grows on south faces the adjacent north faces are often covered with a very much wetter type of forest such as some form of *Schima–Castanopsis* forest or the michelias and laurels of the lower temperate broadleaved forest.

In the East Midlands this type of oak forest is still present in small amounts in the lower Arun and Tamur valleys, but in the very wet upper parts of these valleys it is almost completely absent even on south faces. In the Central Midlands it is also absent from the wet southern flanks of Annapurna Himal and Himal Chuli.

The altitudes at which these oaks grow, 4–8,000 ft, coincide with the zone of maximum cultivation, so that very frequently the oak forest exists only in an isolated and much altered form surrounded by fields and deforested hillsides. In the country west of the Karnali and south of Silgarhi the population is much less dense than in most other Midland areas of Nepal and uncut oak forest still exists here over wide areas. Elsewhere the oak forest has been much damaged, and can be seen in its natural form only in a few places on the outlying spurs of the main ranges. Many intermediate forms of degraded oak forest can be found, and there are many deforested hillsides where only a few coppiced oak stumps survive. In all except the most inaccessible parts of the country the oaks are lopped for cattle fodder during the dry months, and even where trees of good height remain the spread of their branches is often much reduced.

In forest on dry south-facing slopes the oaks do not usually exceed 50 ft in height, and are widely spaced. To improve the grazing the forest is frequently burned in the spring, so that it has a very open appearance with few under-shrubs. Only two species are almost always present, *Rhododendron arboreum* and *Lyonia ovalifolia*. Of all the Nepal forests this is perhaps the least interesting for the botanist. The *Rhododendron* has its brief moment of glory in the spring, for at these altitudes its flowers have a richer, deeper shade than this species attains in the upper forest. Thereafter the forest has little to offer either in colour or variety.

Species typical of this dry form of oak forest are as follows:

1 Quercus incana, Quercus lanuginosa.

2 Rhododendron arboreum, Lyonia ovalifolia, Rhus wallichii.

3 Inula cappa and species of Rubus, Berberis, Leptodermis, and Sarcococca.

In places which are wetter or less frequently burnt the forest is considerably richer in species. The oaks may be 80 ft tall and draped with lichen and epiphytic orchids and ferns. The understory is variable but sometimes can be dense. In one or two places west of the Karnali I have seen the tall palm *Trachycarpus tekel* growing in oak forest at about 7,000 ft. It looks very incongruous in an otherwise entirely temperate forest.

In the West Midlands I have noted the following species growing in this richer type of oak forest:

1 Quercus incana, Quercus lanuginosa.

2 Carpinus viminea, Ilex dipyrena, Symplocos crataegoides, Cornus capitata, Myrica esculenta, Lindera pulcherrima, Neolitsea umbrosa.

3 Lonicera quinquelocularis, Rhamnus virgatus, Viburnum coriaceum, Viburnum stellulatum, Buddleja paniculata, Myrsine semiserrata, Cocculus laurifolius, Gaultheria fragrantissima, Cotoneaster microphylla, Reinwartia trigyna, Myrsine africana and species of Rubus and Berberis.

4 Clematis montana, Clematis connata, Scurrula elata, Jasminum officinale, and species of Sabia, Smilax, and Vitis.

Other species of oak overlap with *Quercus incana* and *Quercus lanuginosa* in some places. *Quercus glauca* is often present in damper localities, and above 7,000 ft *Quercus semecarpifolia* begins to predominate. In the West Midlands *Quercus dilatata* may also be present, but in the Central and East Midlands this species is replaced by *Quercus lamellosa*. In places between 7–8,000 ft a mixture of several of these species can sometimes be found. Where the soil is deep and damp these mixed oaks may form a tall dark forest with a close canopy and few under-shrubs beneath. This forest is not typical of *Quercus incana–Quercus lanuginosa* forest, nor is either species usually dominant in it.

QUERCUS DILATATA FOREST

Champion: Quercus dilatata-Acer forest, p. 233.

In Nepal several species of *Acer* commonly occur with *Quercus dilatata*, but so also do a number of other broadleaved trees. In some places here it is difficult to draw a dividing line between *Quercus dilatata* forest and *Aesculus–Juglans–Acer* forest. Champion refers to the latter type of forest as mixed temperate deciduous forest. *Quercus dilatata* Lindl. is also called *Quercus floribunda* A. Camus. The former name, whether correct or not, appears to be the one most widely known at the present time.

Quercus dilatata is a West Himalayan species with a range extending from Afghanistan to Nepal. Rather surprisingly a variety of this species has also been recorded from localities far away in Western China, although it appears to be absent from the Eastern Himalaya to the east of Nepal. (See the section dealing with distribution, p. 150.)

This oak is fairly common in the West Midlands and occurs in small numbers in the Humla–Jumla area. I have not seen it east of the Kali Gandaki, but J. Kawakita reports it from the Shiar khola.[11]

Quercus dilatata is a tall tree up to 100 ft in height, and in Nepal it occurs between 7–9,500 ft. At these altitudes much of the West Midlands are covered with *Quercus incana–Quercus lanuginosa* forest or with *Quercus semecarpifolia* forest. *Quercus dilatata* is mostly confined to north or west faces where the soil is damp.

I have not seen *Quercus dilatata* forming pure forest over areas of any great extent, and perhaps in Nepal it hardly merits mention as a separate forest type. It does, however, quite commonly occur mixed with *Aesculus indica*, *Ilex dipyrena*, *Alnus nepalensis*, *Juglans regia*, and several species of *Acer* to form a forest much more diverse and interesting than true oak forest. In this mixed forest the oak may be dominant in height, but not in numbers.

A few trees of *Quercus incana* and *Quercus semecarpifolia* are often also present in this mixed forest. There is, however, another minor type of forest found occasionally on these damp slopes at about 7,500 ft in which these two oaks combine with *Quercus dilatata* to form a tall dark forest with a close canopy composed almost exclusively of oak. Here the undergrowth is often scanty, but sometimes there is a dense understory of small moss-clad trees such as *Symplocos ramosissima*, *Neolitsea umbrosa*, *Lindera pulcherrima*, and *Dodecadenia grandiflora*. Where this is so the forest has an appearance very similar to *Quercus lamellosa* forest found at similar altitudes in the Central and East Midlands, except that the numerous epiphytic ericaceous species found in the latter type of forest are absent here.

In the very mixed type of forest found in the West Midlands on north slopes between 7–9,500 ft, where *Quercus dilatata* is dominant in small areas and scattered throughout, I have noted the following species:

1. Quercus dilatata, Quercus incana, Quercus semecarpifolia, Tsuga dumosa, Abies pindrow, Betula alnoides, Alnus nepalensis, Carpinus viminea, Ilex dipyrena, Machilus duthiei, Acer cappadocicum, Acer caesium, Acer sterculiaceum, Acer acuminatum.

2. Symplocos theaefolia, Symplocos ramosissima, Symplocos crataegoides, Neolitsea umbrosa, Lindera pulcherrima, Dodecadenia grandiflora, Rhododendron arboreum, Lyonia ovalifolia, Sorbus cuspidata, Prunus cornuta, Prunus carmesina, Rhus semialata, Rhus succedanea, Cornus macrophylla, Cornus capitata, Rhamnus purpureus, Meliosma dilleniifolia, and species of Taxus.

3. Viburnum stellulatum, Lonicera quinquelocularis, Lonicera webbiana, Daphne bholua, Rosa macrophylla, Ribes acuminatum, Jasminum humile, Mahonia napaulensis, Acanthopanax cissifolius, Rhus cotinus, Staphylea emodi, Pentapanax leschenaultii, and species of Berberis, Sarcococca, and Arundinaria.

4. Euonymus echinatus, Hedera nepalensis, Sabia campanulata, Schizandra grandiflora, Hydrangea anomala, Aristolochia griffithii, Holboellia latifolia, and species of Vitis.

QUERCUS SEMECARPIFOLIA FOREST

Champion: Quercus semecarpifolia forest, p. 235.

Quercus semecarpifolia, the prickly-leaved oak, ranges from Afghanistan through the NW. Himalaya, Nepal, and Bhutan to Manipur. It is reported to be absent from Sikkim.[12] In many parts of Nepal it is widespread at altitudes above 7,500 ft.

Humla–Jumla area

Here much of the forest above 10,000 ft consists of *Abies spectabilis, Quercus semecarpifolia,* and *Betula utilis.* In some places *Abies* and *Quercus* grow together, with the oak forming an understory beneath the fir; I refer to this type of forest in the section dealing with *Abies spectabilis* forest. Very often, however, birch, fir, and oak separate out into almost pure stands, and where this is so the oak occupies the sunny south faces, the fir the north faces, and the birch the gulleys and steep slopes where the snow lies late into the spring. In the pure oak forest the trees may be up to 100 ft tall, and quite frequently they abut directly onto open alpine slopes without any intervening *Betula* forest. Sometimes there is a dense growth of bamboo beneath the oaks, but more often the forest is open and one may be able to get an unimpeded view of a hundred yards or more between the trees.

In this type of forest I have noted the following species:

1 Quercus semecarpifolia. A few Abies spectabilis, Betula utilis.

2 Rhododendron arboreum, Sorbus foliolosa, Taxus species.

3 Rosa sericea, Cotoneaster acuminata, Lonicera myrtillus, Lonicera purpurascens, Ribes glaciale, Arundinaria species.

Much of the Humla–Jumla area between 7,500–10,500 ft is covered with *Pinus–Picea* forest, or more rarely with *Picea–Abies pindrow* forest. *Quercus semecarpifolia* occurs frequently in both these types of forest, but not as a dominant.

The activities of man have very much modified the original forest cover here, and on the fringe of cleared areas one often finds dense coppices of pure *Quercus semecarpifolia,* sometimes accompanied by *Juniperus wallichiana.* Around Jumla the population is still sufficiently small in relation to the size of the country it occupies to be able to practice a limited amount of shifting cultivation. It is in *Quercus semecarpifolia* forest that this shifting cultivation takes place. Small areas of marginal woodland are felled and burnt, and buckwheat and potatoes are cultivated for a few years amongst the stumps before the ground once again reverts to forest.

In other parts of Nepal it is common enough to see crops planted amongst trees which have been newly felled and burnt, but usually this is the first stage towards a permanent clearance and the establishment of a new village.

Quercus semecarpifolia becomes rare as one ascends the Humla Karnali towards Simikot and it is absent from the Mugu khola. Both here and in other valleys which lead up towards the drier country along the Tibetan border I have noticed that one of the first signs of transition to a dry forest flora is the disappearance of the oak.

The West Midlands

Very often the *Quercus incana–Quercus lanuginosa* forest which is ubiquitous in the West Midlands between 4,500–8,000 ft is succeeded above that height by *Quercus semecarpifolia* forest. Between 8,000–10,000 ft, if the slope be a damp north-facing one, *Tsuga* forest, *Quercus dilatata* forest or *Aesculus–Juglans–Acer* forest may occur, but on most other slopes at these altitudes and particularly on dry south-facing ones *Quercus semecarpifolia* forest is widespread.

This oak forest varies much in composition. On dry slopes where burning takes place the trees are often stunted and the composition of the forest is very poor. In moister localities there may be fine tall trees and a much richer composition.

I take an example of the former type from the country north of Dhorpatan. Here on south-facing limestone slopes between 9,500–11,000 ft the oak trees do not exceed 40 ft in height and are widely spaced with much grass beneath them. Fire appears to be frequent and in consequence there are few undershrubs. Some *Pinus excelsa* grows amongst the oak, and doubtless its presence is due to the damaged state of the forest, for it is not a typical component of this oak forest in its natural state. In this locality the forest composition is as follows:

1 Quercus semecarpifolia, Pinus excelsa.

2 Rhododendron arboreum, Lyonia ovalifolia.

3 Jasminum humile, Viburnum cotinifolium, Cotoneaster microphylla, Rhododendron lepidotum (both this species and R. arboreum grow here on limestone rocks) and species of Indigofera and Berberis.

In other places where the soil is moist and deep, *Quercus semecarpifolia* attains a height of 100 ft and forms an almost pure top canopy. In the West Midlands I have noted the following species in this big oak forest:

1 Quercus semecarpifolia.

2 Ilex dipyrena, Acer sterculiaceum, Acer pectinatum, Neolitsea umbrosa, Dodecadenia grandiflora, Lindera pulcherrima, Sorbus cuspidata, Prunus cornuta, Rhododendron arboreum.

3 Ribes glaciale, Viburnum cotinifolium, Viburnum grandiflorum, Rhamnus purpureus, Daphne papyracea, Lonicera myrtillus, and species of Berberis and Sarcococca.

4 Hedera nepalensis, Holboellia latifolia, Clematis montana, Euonymus echinatus.

In one or two places in the West Midlands I have seen a dense understory of *Neolitsea umbrosa, Lindera pulcherrima*, and *Dodecadenia grandiflora* beneath the oak, but this dense growth of laurels is exceptional here and more typical of the Central or East Midlands.

Above 10,000 ft much of the forest in the West Midlands as well as in the Humla–Jumla area consists of *Quercus semecarpifolia, Abies spectabilis*, and *Betula utilis*. The oak grows both mixed with the fir and also in separate stands in the manner I have described for that area.

Central and East Midlands

Here the habitats in which *Quercus semecarpifolia* occurs are more limited than in the west. On north faces it often seems unable to compete successfully with other East Himalayan types of forest, and here it may be replaced by *Quercus lamellosa, Quercus lineata*, and in the extreme east by *Lithocarpus pachyphylla* as well. It is almost completely absent from areas of very heavy rainfall such as the upper Arun and Tamur and the hills north of Pokhara.

In most parts of the area, however, *Quercus semecarpifolia* remains quite common on south faces and on the sides of the big river valleys, very often succeeding *Quercus lanuginosa* forest at about 8,000 ft and being succeeded by *Abies spectabilis* forest at about 10,000 ft. It is far less abundant in and very often absent from the forest above this altitude; the *Abies* forest here is not broken up by stands of pure *Quercus semecarpifolia* reaching to the treeline as it often is in the west.

On slopes which are burnt the composition of the oak forest may be as poor as it is in similar places in the West Midlands, but elsewhere it is enriched by a number of East Himalayan species such as *Edgeworthia gardneri, Leucosceptrum canum,* and *Rhododendron dalhousiae,* and the trees are often densely covered by mosses, ferns, and epiphytes. *Quercus lamellosa* may also be present. Sometimes there may be an overlap between *Quercus semecarpifolia* forest and upper temperate broadleaved forest, so that the oak may be found growing with trees such as *Magnolia campbellii, Michelia doltsopa, Osmanthus suavis,* or *Acer campbellii.* But in the most typical form of *Quercus semecarpifolia* forest both here and in the West Midlands the oak is very dominant in the upper canopy.

CASTANOPSIS TRIBULOIDES–CASTANOPSIS HYSTRIX FOREST

Champion: No separate mention.

Both the above species of *Castanopsis* are East Himalayan in distribution, but whereas *Castanopsis tribuloides* extends its range westwards throughout Nepal and into Kumaon, *Castanopsis hystrix* has not as yet been recorded west of Okhaldunga in the Central Midlands. The spiny involucres which enclose the nuts of these two species are respectively very distinct; their leaves are less readily distinguishable.

In Sikkim the altitudes 6–9,500 ft are often covered with one continuous belt of big, dark, evergreen forest, the lower part of which is composed predominately of *Castanopsis tribuloides* and *Castanopsis hystrix,* the middle part of *Quercus lamellosa,* and the upper part of *Lithocarpus pachyphylla.* In Nepal this evergreen belt is usually limited to *Quercus lamellosa,* for *Lithocarpus pachyphylla* occurs only in a few places, and the *Castanopsis* belt of the lower part which at one time must have been quite widespread has very largely been destroyed to make way for cultivation. Often the *Castanopsis* now occurs here only as thickets and shrubberies surrounding the topmost villages.

There are, however, a few places in the East Midlands at altitudes between 6–7,000 ft where almost pure *Castanopsis* forest survives. This forest is tall and dark with trees of 80–100 ft forming a close canopy; the understory beneath is composed largely of species of *Symplocos* and of the family *Lauraceae. Castanopsis indica* and *Schima wallichii* are absent, and the forest is temperate rather than subtropical in character. In general appearance it is very similar to *Quercus lamellosa* forest.

In contrast with *Castanopsis indica* which flowers in the autumn, *Castanopsis tribuloides* and *Castanopsis hystrix* flower in the spring. The fresh yellow leaves of lopped trees of *Castanopsis tribuloides* are very prominent in April in the shrubberies which surround the villages of the Arun and Tamur.

In *Castanopsis* forest I have noted the following species:

1 Castanopsis tribuloides, Castanopsis hystrix, Quercus lamellosa.

2 Lindera pulcherrima, Neolitsea umbrosa, Neolitsea lanuginosa, Litsea elongata, Machilus odoratissima, Symplocos sp., Ilex insignis, Michelia velutina, Macropanax oreophilum, Rhododendron arboreum, Leucosceptrum canum, Eurya acuminata.

3 Mahonia napaulensis, Viburnum erubescens, Viburnum coriaceum, Edgeworthia gardneri, Daphne papyracea, Cornus oblonga, and species of Sarcococca, Berberis and bamboo.

4 Vaccinium retusum, Vaccinium vaccinaceum, Agapetes serpens, Hymenopogon parasiticus, Hedera nepalensis, Lonicera glabrata, epiphytic ferns and orchids.

QUERCUS LAMELLOSA FOREST

Champion: E. Himalayan wet temperate forest; subtype buk oak forest, p. 221.

Quercus lamellosa occurs in the Eastern Himalaya and in Western China. It is absent from the Western Himalaya. I myself have not seen this species west of the Kali Gandaki, but it is said to grow on wet north faces to the south of Dhaulagiri.

There is another species of oak commonly found in *Quercus lamellosa* forest to which reference must be made and about the specific status of which there seems to be some uncertainty. The acorn cups of this oak are formed of concentric rings in the same manner as those of *Quercus lamellosa*, but both the acorns and the leaves are much smaller than those of this latter species. In the past this oak has usually been referred to as *Quercus lineata* or one of its varieties. I am informed, however, that the true *Quercus lineata* is confined to the Malaysian area, and does not occur in the Himalaya at all, and that this oak should correctly be called *Quercus oxyodon*. In view, however, of the fact that this tree is known to most people who are familiar with it in the field as *Quercus lineata* I shall continue to call it so in these notes.

There is a further point of doubt about this oak. G. King stated his belief 'that *Quercus glauca* and *Quercus lineata* are really but forms of one widely distributed species which is found from Japan to Java and runs westward along the Himalaya as far as Hazara.'[13] Hiroo Kanai also writes of the gatherings of *Quercus lineata* made by the Japanese expeditions in the Eastern Himalaya. 'The size and shape of the leaf and acorns and the amount of hairiness of leaf vary greatly. There are many intermediate forms between this and *Quercus glauca*.'[14] It seems, therefore, that further taxonomic work must be done before the problems surrounding this oak are cleared up.

Quercus lamellosa forest is particularly abundant on the ridges that flank the upper Arun and Tamur, and on the southern slopes of Himal Chuli and Annapurna Himal. Here it forms a more or less continuous belt, coming in above *Castanopsis* forest or the laurels and michelias of the lower temperate mixed broadleaved forest, and running up into the magnolias and maples of the upper temperate mixed broadleaved forest. The altitudes at which it occurs lie between 6,500–8,500 ft.

In other drier parts of the Central and East Midlands *Quercus lamellosa* is much less widespread, and here usually it is confined to north or west faces. In localities which probably are approaching the minimum of moisture which the species will tolerate it sometimes occurs sandwiched in between *Quercus lanuginosa* below and *Quercus semecarpifolia* above, all three species mingling together to form a mixed oak forest.

In its purest form *Quercus lamellosa* forest is somewhat dark and gloomy. In areas of very heavy rainfall where the forest is seen at its best *Quercus incana*, *Quercus lanuginosa* and *Quercus semecarpifolia* are altogether absent. The upper canopy is formed very largely of *Quercus lamellosa*, with small quantities of *Quercus lineata* and perhaps a few *Castanopsis tribuloides* in the lower part. In the forests of the upper Arun two other tree species are very often present, *Litsea elongata* and *Ilex sikkimensis*, and here the forest tends to be fairly open beneath. Elsewhere there is usually a dense understory of smaller trees, of which the most prominent are species of *Lauraceae* and *Symplocos*. The trees are draped with mosses, ferns, and epiphytes, and the understory forms dense thickets through which visibility is restricted and movement difficult. Species of *Mahonia*, *Sarococcoca*, and *Skimmia* are common here, and *Cymbidium grandiflorum* is prominent when flowering in the spring. *Rhododendron dalhousiae* grows both on rocks in the forest and on the oak trees, and other epiphytic ericaceous species such as *Vaccinium retusum*, *Vaccinium nummularia*, and *Agapetes serpens* are common. There may be a little *Rhododendron arboreum* present, but other species of this genus are normally absent. Some of the laurel trees in this type of forest may attain a height of 70 ft.

The upper limit of cultivation in the wet parts where *Quercus lamellosa* grows is at about 7,000 ft, so that frequently the lower parts of the *Quercus lamellosa* forest have been cleared. In these places one enters the forest through the much-lopped and leech-infested shrubberies which fringe the topmost fields. Where the oak has been cut secondary forest consisting almost entirely of species of *Symplocos* and of the family *Lauraceae* may occur, and occasionally *Daphniphyllum himalayense* forms almost pure secondary forest at these altitudes.

In the Central and East Midlands I have noted the following species occurring in *Quercus lamellosa* forest:

1 Quercus lamellosa, Quercus lineata, Castanopsis tribuloides.

2 Ilex sikkimensis, Ilex dipyrena, Litsea elongata, Machilus duthiei, Machilus odoratissima, Dodecadenia grandiflora, Neolitsea umbrosa, Lindera pulcherrima, Symplocos ramosissima, Symplocos theaefolia, Symplocos sumuntia, Michelia doltsopa, Acer sterculiaceum, Acer hookeri, Maytenus rufa, Lyonia ovalifolia, Rhododendron arboreum, Schefflera impressa, Brassaiopsis glomerulata, Daphniphyllum himalayense, Prunus nepalensis.

3 Mahonia napaulensis, Edgeworthia gardneri, Helwingia himalaica, Viburnum erubescens, Brassaiopsis mitis, Skimmia arborescens, and species of Sarocococca and Arundinaria.

4 Rhododendron dalhousiae, Rhododendron lindleyi, Vaccinium retusum, Vaccinium nummularia, Agapetes serpens, Hedera nepalensis, Euonymus echinatus, Hydrangea anomala, Hymenopogon parasiticus, Cymbidium grandiflorum.

LITHOCARPUS PACHYPHYLLA FOREST

Champion: E. Himalayan wet temperate forest; subtype high level oak forest, p. 221.

Lithocarpus pachyphylla is an East Himalayan species. In Nepal I have seen it only in the extreme east of the country, in the part which lies between the Tamur and the Sikkim border. In some places here it predominates between 8–9,500 ft.

In its lower parts *Lithocarpus pachyphylla* forest merges with *Quercus lamellosa* forest, and there may also be a number of trees of *Quercus lineata* present. In its upper part it merges with *Rhododendron* forest or with the maples and magnolias of the upper temperate mixed broadleaved forest. In its most typical form, however, the upper canopy of this forest is composed almost exclusively of *Lithocarpus*, the trees attaining a height of 80–100 ft. The canopy is dense and the forest dark, and species of *Symplocos* and of the family *Lauraceae* predominate in the understory. In the spring the white flowers of *Rhododendron grande*, *Magnolia campbellii*, and *Michelia doltsopa* are prominent amongst the evergreen foliage.

In *Lithocarpus pachyphylla* forest I have noted the following species:

1 Lithocarpus pachyphylla, Quercus lamellosa, Quercus lineata.

2 Ilex dipyrena, Ilex sikkimensis, Ilex hookeri, Magnolia campbellii, Michelia doltsopa, Osmanthus suavis, Schefflera impressa, Acer campbellii, Rhododendron grande, Rhododendron falconeri, Neolitsea umbrosa, Litsea elongata, Litsea kingii, Litsea sericea, Lindera pulcherrima, Dodecadenia grandiflora, Symplocos theaefolia, Symplocos dryophila, and species of Prunus and Taxus.

3 Daphne bholua, Ribes glaciale, Viburnum cordifolium, Berberis insigne, Sarcococca species.

4 Vaccinium nummularia, Vaccinium retusum.

AESCULUS–JUGLANS–ACER FOREST

Champion: Moist temperate deciduous forest, p. 257.

This is a West Himalayan type of forest which occurs in the West Midlands, the Humla–Jumla area, and in some of the inner valleys of the western half of Nepal. In more easterly parts of the country this type of forest is replaced by upper temperate mixed broadleaved forest containing a number of East Himalayan species such as *Magnolia campbellii, Acer campbellii,* and *Osmanthus suavis,* which have not been recorded from West Nepal.

Aesculus indica is perhaps most typical of the trees composing the forest. This West Himalayan species is fairly common in the West Midlands and the Humla–Jumla area, but it disappears in the Pokhara district and I have not seen it further east. *Acer caesium,* another West Himalayan species, is the most common maple in the forest. I know of no record of this species occurring east of the Kali Gandaki. *Juglans regia* is also a prominent component of the forest. This species is much more widespread than the two former ones, occurring throughout the Himalaya. In West Nepal it is common as a village tree, but it is not grown as a village tree in eastern parts of Nepal, nor is it nearly so common in the forest there.

The *Aesculus–Juglans–Acer* forest of the West Midlands differs somewhat from that of the Humla–Jumla area, and I describe them separately.

The West Midlands

Aesculus–Juglans–Acer forest often occurs here between 6–9,000 ft in quite narrow strips on flat terraces alongside streams. In addition to the three species mentioned above *Betula alnoides* and *Alnus nepalensis* are common in these waterside places, and a few trees of *Quercus semecarpifolia, Quercus incana,* or *Quercus dilatata* may be present.

These flat places are much used for grazing and the forest frequently consists of tall mature trees 100 ft in height with a close canopy and very little undergrowth beneath. In less grazed forest, however, undergrowth may be dense. The chestnuts are a fine sight when they flower in May.

In this type of forest in the West Midlands I have noted the following species:

1 Aesculus indica, Juglans regia, Acer caesium, Betula alnoides, Alnus nepalensis. A few Quercus dilatata, Quercus semecarpifolia, Quercus incana.

93

2 Populus ciliata, Ilex dipyrena, Acer sterculiaceum, Prunus cornuta, Pyrus pashia, Machilus duthiei, Lindera pulcherrima, Neolitsea umbrosa.

3 Prinsepia utilis, Rosa brunonii, Coriaria nepalensis, Spiraea sorbifolia, Rhamnus purpureus, Philadelphus tomentosus, and species of Sarocococca.

4 Hedera nepalensis, Hydrangea anomala.

Aesculus–Juglans–Acer forest also occurs away from these river flats, but here it is less easily distinguishable from other forest types. In the West Midlands north-facing slopes between 6–9,000 ft often are covered with a very mixed forest. The bottom part between 6–7,000 ft may contain some michelias and numerous laurels; between 7–9,000 ft there may be *Quercus dilatata* mixed with deciduous broadleaved trees; in places there may be *Tsuga dumosa;* and west of the Karnali *Abies pindrow* may be added to the mixture. *Aesculus indica, Juglans regia* and *Acer caesium* may occur anywhere in this mixture, and in some places they may be locally dominant.

I have attempted to describe this very mixed forest in the section dealing with *Quercus dilatata* forest.

The Humla–Jumla Area

Aesculus–Juglans–Acer forest occurs here between 6,500–9,500 ft. Most of the forest at these altitudes is *Pinus–Picea* forest, and the broadleaved forest is confined either to flat streamside sites similar to those described for the West Midlands or to damp gulleys in the conifer forest. The broadleaved forest is quite widespread, but never very extensive.

The *Aesculus–Juglans–Acer* forest here differs in a number of ways from that found in the West Midlands. Oak is only occasionally present, and instead a few conifers commonly mingle with the broadleaved trees. *Populus ciliata* largely replaces *Alnus nepalensis* and *Betula alnoides* along the streams. *Betula utilis* is common even as low as 9,000 ft, and laurels are absent. A number of West Himalayan species occur in this forest in the Humla–Jumla area: *Corylus colurna, Ulmus wallichiana, Morus serrata, Acer cappadocicum, Picrasma quassioides.*

On the flats the broadleaved trees may be tall, but in the gulleys running up into the conifer forest they do not much exceed 30–40 ft. Of all the Nepal forests this Humla–Jumla broadleaved forest ressembles most closely a European woodland; chestnut, hazel, poplar, walnut, birch, willow, maple, box, and yew give it a very homely appearance.

In the Humla–Jumla area I have noted the following species in this type of forest:

1 Aesculus indica, Juglans regia, Ulmus wallichiana, Populus ciliata, Betula utilis, Acer caesium, Acer cappadocicum, Acer sterculiaceum, Acer acuminatum, Prunus cornuta; a few Abies pindrow, Pinus excelsa, Picea smithiana; rarely Betula alnoides, Alnus nepalensis, Quercus dilatata, Quercus semecarpifolia.

2 Euonymus tingens, Euonymus fimbriatus, Corylus colurna, Taxus sp., Cornus macrophylla, Cornus capitata, Rhus wallichii, Rhus succedanea, Morus serrata, Picrasma quassioides, Meliosma dilleniifolia, Hydrangea heteromalla, Salix species.

3 Viburnum stellulatum, Staphylea emodi, Syringa emodi, Rhus cotinus, Spiraea sorbifolia, Deutzia hookeriana, Buxus sempervirens, and species of Ligustrum, Sarcococca, Berberis, and Arundinaria.

LOWER TEMPERATE MIXED BROADLEAVED FOREST

Champion: E. Himalayan wet temperate forest, p. 221.
Champion distinguishes three altitudinal zones in this forest: (*a*) laurel forest, (*b*) *Quercus lamellosa* forest, (*c*) *Quercus pachyphylla* forest. The two latter types I describe elsewhere; laurel forest I include in lower temperate mixed broadleaved forest.

Lower temperate mixed broadleaved forest is found in the wetter parts of Nepal between 5–7,000 ft, usually on north or west faces. At these altitudes there are many fields and villages, and in consequence much of this forest has been destroyed. The forest is best seen in the side valleys of the upper Arun and Tamur, and in the wet country to the south of Himal Chuli and Annapurna Himal, but it also occurs in small patches in damp places throughout the Midland areas. I have seen forest of this type as far west as the southern side of the chain of lekhs which separate Jumla from the Midlands.

The forest is to a large extent evergreen, and in many places trees of the family *Lauraceae* are prominent. *Machilus duthiei, Machilus odoratissima, Machilus sericea, Phoebe lanceolata, Phoebe pallida, Cinnamomum tamala, Actinodaphne reticulata, Lindera bifaria, Lindera neesiana, Litsea oblonga, Litsea citrata, Neolitsea umbrosa,* and *Neolitsea lanuginosa* can be all found in this type of forest, though of

95

all these species only *Machilus duthiei, Machilus odoratissima, Neolitsea umbrosa,* and *Cinnamomum tamala* are common.

Because of the abundance of laurels some authorities have described this forest as it occurs in Sikkim as 'laurel forest' or as 'michelia–laurel forest'. It is tempting to apply some such brief expressive title to this forest as it occurs in Nepal, but I refrain from doing so for the following reasons:

1 Laurels in Nepal are by no means confined to this type of forest. As well as being quite numerous in places in the subtropical forest they tend to form a belt in wet localities at altitudes between 5–8,000 ft or even higher. This laurel belt passes through several different types of forest; *Quercus lamellosa* forest, *Aesculus–Juglans–Acer* forest, and the wetter kind of *Quercus incana* forest. To name the forest at one particular altitude 'laurel forest' would be to ignore the wide spread of this belt.

2 Although laurel species such as *Machilus duthiei, Machilus odoratissima* and *Cinnamomum tamala* can be tall trees which reach the upper canopy, many of the species listed only occur in the understory. It would be as unreasonable to call such forest 'laurel forest' as it would be to call all the different types of forest in which rhododendrons are numerous 'rhododendron forest'.

3 In michelia–laurel forest as recorded from Sikkim the *Michelia* species referred to is *Michelia (Alcimandra) cathcartii.* This species has not as yet been recorded from Nepal. The tree which occurs at these altitudes in Nepal is *Michelia kisopa.* Usually this species occurs in a very mixed type of forest, and only occasionally does it dominate the forest over very limited areas.

4 This mixed character of the forest must be stressed. Trees such as *Lithocarpus spicata, Castanopsis tribuloides, Quercus glauca, Carpinus viminea, Eriobotrya elliptica, Turpinia nepalensis,* and *Evodia fraxinifolia* are as common as the *Michelia,* and may occur with or without laurels beneath them.

5 In the lower parts of the forest there is also a strong element of mixed subtropical species, such as *Acer oblongum, Albizzia mollis, Engelhardtia spicata, Zizyphus incurva,* and *Homalium nepalense.*

6 At the densely populated altitudes at which this forest occurs the forest composition has often been altered by the addition of species which are secondary in origin, such as *Leucosceptrum canum, Ehretia macrophylla, Mallotus nepalensis,* and *Daphniphyllum himalayense.*

This latter point underlines the difficulties of classifying this type of forest in Nepal. If the forest was more extensive and less damaged the problem would be easier.

One feature which distinguishes this forest from temperate forests at higher altitudes is the number of species which flower in autumn or early winter. Examples are *Michelia kisopa*, *Alnus nepalensis*, *Homalium nepalense*, *Symplocos theaefolia*, *Leucosceptrum canum*, *Sarauja napaulensis*, *Prunus undulata*, *Eurya acuminata*, *Cornus oblonga*, *Camellia kissi*, and a number of species of the families *Lauraceae* and *Araliaceae*.

In the West Midlands this type of forest occurs only rarely, mostly in shady gulleys in *Quercus incana* forest. Between 5–7,000 ft on the south side of the lekhs which separate Jumla from the Midlands I have noted the following species:

1 Michelia kisopa, Lithocarpus spicata, Quercus glauca, Castanopsis tribuloides, Machilus duthiei, Machilus odoratissima, Cinnamomum tamala, Betula alnoides, Alnus nepalensis, Dalbergia hircina, Albizzia mollis, Acer oblongum, Cedrela species.

2 Lindera pulcherrima, Neolitsea umbrosa, Cinnomomum species, Dodecadenia grandiflora, Eriobotrya elliptica, Sapium insigne, Daphniphyllum himalayense, Macaranga denticulata, Myrsine semiserrata, Pyrularia edulis, Brassaiopsis aculeata, Meliosma dilleniifolia, Rhus succedanea.

3 Maesa chisia, Daphne papyracea, Mahonia napaulensis, Cornus oblonga.

4 Hydrangea anomala, Hedera nepalensis, Euonymus echinatus, and species of Rhaphidophora and Sabia. Epiphytic ferns and orchids common.

I have not seen this type of forest in the Humla–Jumla area. Here laurels are conspicuously absent, though at one place at 7,000 ft in the Mugu Karnali valley there is a rather surprising little wood composed of *Quercus incana*, *Machilus duthiei*, *Betula alnoides*, and *Alnus nitida* with a dense understory of *Neolitsea umbrosa*.

In the Central and East Midlands the lower temperate mixed broadleaved forest is found more frequently. Here a belt of *Quercus lamellosa* forest very often separates its upper part from the upper temperate mixed broadleaved forest. In its lower part it may mingle with the tree ferns and *Pandanus* of the subtropical forest, or with *Castanopsis indica* and *Castanopsis tribuloides*.

The composition of the forest in the Central and East Midlands is very much richer than in the West Midlands, and includes a number of East Himalayan species. I have noted the following species:

HFN   97

1 Lithocarpus spicata, Quercus glauca, Castanopsis tribuloides, Machilus duthiei, Machilus odoratissima, Neolitsea lanuginosa, Litsea oblonga, Phoebe lanceolata, Phoebe pallida, Michelia kisopa, Alnus nepalensis, Betula alnoides, Juglans regia, Acer oblongum, Ehretia macrophylla, Engelhardtia spicata, Schima wallichii, Michelia doltsopa, Bucklandia populnea, Carpinus viminea, Acer thomsonii.

2 Rhus succedanea, Rhus semialata, Turpinia nepalensis, Eriobotrya elliptica, Zizyphus incurva, Homalium nepalense, Acer sikkimensis, Mallotus nepalensis, Daphniphyllum himalayense, Symplocos theaefolia, Symplocos ramosissima, Prunus undulata, Evodia fraxinifolia, Leucosceptrum canum, Rhododendron arboreum, Sarauja napaulensis, Meliosma pungens, Talauma hodgsonii, Prunus nepalensis, Macropanax oreophilus, Macaranga pustulata.

3 Viburnum stellulatum, Viburnum coriaceum, Viburnum erubescens, Brassaiopsis hainla, Brassaiopsis aculeata, Mezoneurum cucullatum, Lindera bifaria, Lindera neesiana, Actinodaphne reticulata, Cornus oblonga, Camellia kissi, Boehmeria platyphylla, Boehmeria macrophylla, Ardisia macrocarpa, Maesa chisia, Pseuderanthemum indicum, and species of Mussaenda, Sarcococca, and Arundinaria.

4 Schefflera venulosa, Hedera nepalensis, Jasminum dispermum, Photinia integrifolia, Euonymus echinatus, Thunbergia coccinea, Vaccinium dunalianum, Vaccinium vaccinaceum, Hoya lanceolata, Hoya fusca, and species of Aeschynanthus and Raphidophora.

UPPER TEMPERATE MIXED BROADLEAVED FOREST

Champion: Under the heading 'Moist temperate deciduous forest' (p. 258) Champion describes for the Western Himalaya a type of forest very similar to that which I have described above for Nepal under the heading '*Aesculus–Juglans–Acer* forest'. He adds a note to his forest type: 'In the E. Himalaya patches of predominantly deciduous forest occur in the wet temperate forest which are obviously closely related to the western form here described. Further study is needed.'

Although the forest described below is by no means entirely deciduous I think it approximates to the eastern predominantly deciduous forest mentioned by Champion.

A number of the species which compose the *Aesculus–Juglans–Acer* forests of the West Midlands and the Humla–Jumla area do not occur in the wetter Central and East Midlands, though some of them persist for some way eastwards in the inner valleys. Examples are *Quercus dilatata, Aesculus indica, Populus ciliata, Ulmus wallichiana, Corylus colurna, Staphylea emodi, Syringa emodi,* and *Rhus cotinus*.

They are replaced by East Himalayan species such as *Magnolia campbellii, Acer campbellii, Osmanthus suavis, Schefflera impressa,* and *Corylus ferox*. These latter species are typical of the upper temperate mixed broadleaved forest. The approximate dividing line between the two types of forest is the Kali Gandaki river.

Upper temperate mixed broadleaved forest occurs in the Central and East Midlands between 8–10,500 ft, mostly on slopes which face north and west. At these altitudes slopes which face south and east are often covered with *Quercus semecarpifolia,* but this oak is not a normal component of the forest type. *Quercus lamellosa,* however, is often present in the lower part of the forest.

The upper temperate mixed broadleaved forest is predominantly broadleaved and rather less predominantly deciduous. Some trees of *Tsuga dumosa* often occur in it, and these tend to separate out into pure stands on the drier ground. In some places the forest trees are 60–80 ft tall with beneath them a dense under-story of moss-draped smaller trees in which species of *Symplocos, Litsea,* and *Lindera* are common. In other places the broadleaved trees are much smaller and the forest rather open. *Acer* species and *Rhododendron arboreum* are prominent throughout, and *Alnus nepalensis* frequently occurs in wet gulleys in the lower part of the forest. *Vaccinium nummalaria* and *Vaccinium retusum* are the most prominent of the numerous epiphytes.

In this type of forest I have noted the following species:

1 Acer campbellii, Acer sterculiaceum, Acer pectinatum, Magnolia campbellii, Osmanthus suavis, Ilex dipyrena, Ilex fragilis, Sorbus cuspidata, Corylus ferox, Alnus nepalensis, Prunus cornuta, Schefflera impressa, Betula utilis, Populus glauca, Symplocos sumuntia, Symplocos ramosissima, Lindera pulcherrima, Lindera heterophylla, Litsea sericea, Dodecadenia grandiflora, Neolitsea um-brosa, Lyonia ovalifolia, Rhododendron arboreum, Leucosceptrum canum, Euonymus tingens, and species of Taxus and Salix. Tsuga dumosa and Quercus lamellosa are often present, but not dominant.

2 Mahonia napaulensis, Viburnum erubescens, Viburnum cordifolium, Viburnum stellulatum, Leycesteria formosa, Piptanthus nepalensis, Pentapanax leschen-aultii, Acanthopanax cissifolius, Hydrangea heteromalla, Ribes acuminatum, Spiraea bella, Philadelphus tomentosus, Rosa sericea, Cotoneaster frigida, Cotoneaster acuminata, Pieris formosa, and species of Berberis, Daphne,

Sarcococca, Rubus and Skimmia. Some of these shrubs are more typical not of the forest itself, but of the numerous clearings found within it.

3 Holboellia latifolia, Schizandra grandiflora, Aristolochia griffithii, Hymeno-pogon parasiticus, Hydrangea anomala, Euonymus echinatus, Clematis montana, Clematis connata, Lonicera acuminata, Vaccinium retusum, Vaccinium nummularia.

In some places *Rhododendron* forest may replace the upper temperate mixed broadleaved forest from 8,500 ft upwards. These two types of forest share many component species, but in their most typical forms they are quite distinct. In upper temperate mixed broadleaved forest the tree species, many of which are deciduous, predominate. These trees are often 60 ft tall. Species of *Rhododendron* may be present in this forest, but they do not predominate even amongst the undershrubs. In *Rhododendron* forest, on the other hand, trees common in the upper temperate mixed broad-leaved forest only occur scattered thinly amongst dense growths of *Rhododendron*, or may be altogether absent. Forest inter-mediate between the two types also occurs.

RHODODENDRON FOREST

Champion   does not refer to *Rhododendron* forest as a separate type. To a certain extent he includes it within birch–rhododendron forest (p. 271) and alpine fir–birch forest (p. 269).

In my opinion the presence in a forest of much *Rhododendron* does not itself indicate that the forest should be classed as *Rhododendron* forest. In Nepal rhododendrons are often very numerous in *Abies* forest, *Tsuga* forest, *Betula* forest, and upper temperate mixed broadleaved forest, but in all these types of forest they are under-shrubs beneath a more or less continuous canopy com-posed principally of the tree species from which the forests take their names.

In the *Rhododendron* forest described below the rhododendrons themselves dominate the upper story, and such other tree species as may be present only occur irregularly scattered amongst them. At lower altitudes these rhododen-drons may be 40 ft tall, and merit the description 'forest'. At higher altitudes they may be no more than 5–12 ft tall, and are better described as 'shrubberies'. (See the section dealing with moist alpine scrub, p. 128.)

Before describing *Rhododendron* forest it may be convenient to list the *Rhododendron* species which have been recorded in Nepal. Some of these are alpine species which are not found in the forest zone. Many of the species listed are confined to the eastern-most parts of the country. The number of species occurring in western parts is reduced to about seven.

In compiling this list I have omitted the varieties of *R. cinnabarinum*, and the numerous varieties *R. arboreum*. I have lumped *R. hypenanthum* with *R. anthopogon*, and *R. aeruginosum* with *R. campanulatum*.

R. anthopogon, R. arboreum, R. barbatum, R. camelliaeflorum, R. campanulatum, R. campylocarpum, R. ciliatum, R. cinnabarinum, R. cowanianum, R. dalhousiae, R. elaeagnoides, R. falconeri, R. fulgens, R. glaucophyllum, R. grande, R. griffithianum, R. hodgsonii, R. lepidotum, R. lindleyi, R. lowndesii, R. nivale, R. pendulum, R. pumilum, R. setosum, R. thomsonii, R. trichocladum, R. triflorum, R. vaccinioides, R. virgatum, R. wallichii.

Rhododendrons in Nepal are most widespread and most numerous in species in areas of very heavy rainfall on the upper Arun and Tamur. In some places here the upper forest between 8,500 ft and the alpine zone is one great sweep of almost pure rhododendron. At lower levels *R. grande*, *R. hodgsonii*, and *R. falconeri* predominate, forming tangled forest 40 ft tall. On ridges and south-facing slopes *R. arboreum* with gnarled and twisted trunks occurs in dense 40-ft thickets. At higher levels *R. campanulatum*, *R. wallichii*, *R. thomsonii*, and *R. campylocarpum* form impenetrable shrubberies 15–20 ft tall. Above this again somewhat smaller shrubberies of *R. fulgens* and *R. wightii* lead into the alpine zone.

This more or less pure *Rhododendron* forest all occurs at altitudes well within the range of other taller forest trees, and one must ask why the taller trees are absent here. One possible reason is that heavy rainfall on steep rocky slopes leaves only a thin layer of soil, which is sufficient for the shallow-rooting rhododendrons but not for bigger tree species.

The above mentioned factor would not control the occurrence at lower altitudes of pure stands of the big-leaved *R. grande*, *R. falconeri*, and *R. hodgsonii*, which tend to occur on flat level ground, and which are often surrounded by slopes covered with conifer forest. In some cases the conifers may have been removed from these flats by felling, but I suspect that the composition of the soil in these places may be partly responsible for inhibiting the growth of all except the rhododendrons.

I know of no publication dealing with the soils of the Himalaya, but a recent comment made by P. J. Grubb on the forests found on tropical mountains perhaps is also relevant to Nepal.[15] He points out that persistent fog can determine the types of forest growing at certain altitudes, because under these conditions the soil water content is raised and the mineralisation of organic matter is slowed down. The increasingly poor supply of nitrogen and phosphorus caused by decrease in temperature and increase in frequency of fog is likely to be important in governing the distribution of forest types. Certainly in Nepal the hillsides on which *Rhododendron* forest predominates over large areas are exceptionally likely to be covered with drizzling mist, both during the monsoon months and also at other times of year.

Pure forest of *R. arboreum* perhaps originates from different causes. Shrubberies composed of this ubiquitous species are common throughout Nepal, particularly on slopes which have been cleared of oak forest. It seems probable that some of the *R. arboreum* forest which occurs on rain-soaked ridges in the eastern parts of the country is of natural origin, but I think that much of such forest now consists almost purely of this species only because other tree species have been felled. The most extensive pure *R. arboreum* forest which I have seen in Nepal occurs on the ridge which runs from Dhankuta northwards to Chainpur.

In *R. arboreum* forest, and to a lesser extent in forest of *R. hodgsonii*, *R. grande*, and *R. falconeri*, associated species of trees and shrubs may be almost entirely absent. At higher altitudes and on ridges where the forest cover is broken other species are a good deal more numerous. Many of these are smaller species of *Rhododendron*, or members of other ericaceous genera. The list of ericaceous species I have noted in these places is as follows:

> Gaultheria hookeri, G. trichophylla, G. nummularioides, G. griffithiana, G. pyrolifolia, Vaccinium retusum, V. nummularia, Lyonia villosa, L. ovalifolia, Pieris formosa, Diplarche multiflora, Enkianthus deflexus.

As well as occasional trees of *Betula utilis*, *Tsuga dumosa*, or *Abies spectabilis*, a number of trees typical of upper temperate mixed broadleaved forest can sometimes be found occurring in *Rhododendron* forest. Apart from the species noted in the section dealing with upper temperate mixed broadleaved forest I have seen the following species in small numbers in *Rhododendron* forest. *Maddenia himalaica, Ilex hookeri, Ilex intricata, Gamblea ciliata, Magnolia globosa, Stachyurus himalaicus, Sorbus thomsonii.*

BETULA UTILIS FOREST

Champion: Alpine fir–birch forest, p. 269. Birch–Rhododendron forest, p. 271.

Betula utilis ranges from Afghanistan to Western China. In Nepal it is widespread and often dominates the forest at the treeline.

The Midlands

In the Midlands *Betula utilis* usually occurs between 11,000–12,500 ft, in many places forming a definite belt at the upper limit of the forest. The typical succession here is from *Abies spectabilis* forest to *Betula* forest to treeless alpine slopes, though sometimes the *Abies* reaches the treeline, and where there are avalanches and landslips the *Betula* penetrates down into the *Abies*.

The upper story of birch forest is composed almost exclusively of birch. The trees rarely exceed 25–30 ft in height, and their gnarled and twisted stems are often covered with peeling strips of pink papery bark. *Rhododendron campanulatum* is the most constant associate of the birch. This type of forest is referred to by Champion as Birch–Rhododendron forest.

In the Central Midlands in birch forest between 11–12,000 ft near the Rupina La I noted the following species:

1 Betula utilis. A few Abies spectabilis.

2 Acer pectinatum, Acer caudatum, Juniperus recurva, Sorbus foliolosa, Prunus rufa, Lyonia villosa, Rhododendron campanulatum, Rhododendron barbatum, Ribes glaciale, Viburnum cordifolium.

3 Clematis montana.

In the East Midlands, and particularly in the upper valleys of the Arun and Tamur, the birch in places seems to be unable to compete with the dense growth of rhododendrons which blanket the country at these altitudes. Often the birch occurs only as individual trees scattered amongst the tangle of rhododendron shrubs, and where it does form forest the undershrubs are almost exclusively ericaceous. In a locality at 12,000 ft in the Hongu khola just to the west of the Arun watershed I noted the following species:

1 Betula utilis.

2 Rhododendron campanulatum, R. fulgens, R. arboreum, R. hodgsonii.

103

The dividing line between this type of birch forest and moist alpine scrub is often indistinct.

In the West Midlands, and particularly in those parts which lie to the west of the Karnali, *Betula utilis* is often found with *Quercus semecarpifolia* and *Abies spectabilis* in associations similar to those I describe for the Humla–Jumla area. This is the type of forest referred to by Champion as Alpine Fir–Birch forest.

The Humla–Jumla area

The birch descends lower here than in the Midlands, and can often be found at 9,000 ft. Between 9–10,000 ft it frequently occurs as a component of the mixed conifer forests of *Pinus excelsa*, *Picea smithiana*, and *Abies pindrow*, and also in the *Aesculus–Juglans–Acer* forest which grows along the streamsides at these altitudes. In these forests the birch is usually a much bigger tree than in the upper forest, attaining heights up to 70 ft.

Between 10–12,500 ft the birch often occurs here mixed with *Abies spectabilis* and *Quercus semecarpifolia*. The relationship of the three species seems to be governed to a large extent by snowfall, and in many places they separate out into almost pure stands; the oak growing on south faces which quickly clear of snow; the fir on north or west faces where the snow lies for longer; and the birch predominating in gulleys and on steep faces where the snow drifts deep in winter and lies until May and even June. Many of the undershrubs of the birch forest are bent sideways by the weight of winter snow.

In some places in the Humla–Jumla area I have seen almost 3,000 ft of hillside covered with birch forest, but this is exceptional and due to the rocky hillside being covered with a thin unstable soil. A patchwork of oak, fir, and birch is more typical of the upper forest here.

Between 10,500–12,000 ft in forest where *Betula* was predominant I noted the following species:

1 Betula utilis.

2 Prunus cornuta, Prunus rufa, Acer caesium, Acer pectinatum, Euonymus porphyreus, Sorbus foliolosa, Sorbus microphylla, Lonicera myrtillus, Lonicera webbiana, Lonicera hispida, Lonicera purpurascens, Ribes emodense, Ribes griffithii, Ribes glaciale, Cotoneaster acuminata, Rhododendron campanulatum and species of Salix and Arundinaria.

3 Lonicera obovata, Rhododendron lepidotum, Cotoneaster microphylla.

4 Clematis montana.

Inner valleys, and edges of the arid zone

In most of the inner valleys of the eastern half of Nepal the significant reduction in rainfall which occurs towards the valley heads takes place at an altitude higher than the treeline. In consequence the birch forest found there differs
little from such forest in the Midland areas.

In the western half of the country the position is different. The valleys which
lead up to Dolpo, Mustang, and Manang show signs of reduced rainfall at much
lower altitudes, and dry forest types such as *Cupressus torulosa, Pinus excelsa,* or
Juniperus wallichiana forest occur here. In these valleys the treeline is much higher
than on the southern sides of the main ranges, and in some places forest can be
found at 14,500 ft.

Birch and pine often form a mixed forest in these places (see the section dealing with *Pinus excelsa* forest, p. 111), but at the highest altitudes most of the
forest consists of pure birch. In such places the hard dry growing conditions and
the presence of the treeless steppes close to the north are reflected in the composition of the forest; dwarf junipers and low *Caragana* shrubs are common and
along the streams grow species of *Salix, Myricaria,* and *Hippophae.* In these
places it is often difficult to distinguish between open birch forest and dry alpine
scrub.

At 12,500 ft in very open birch forest in the Phoksumdo khola on the edge of
Dolpo I noted the following species.

1 Betula utilis. A few Pinus excelsa.
2 Juniperus wallichiana, Juniperus squamata, Caragana gerardiana, Caragana
brevifolia, Lonicera myrtillus, Lonicera spinosa, Rosa sericea, Rosa macrophylla, Potentilla fruticosa, Spiraea arcuata, Abelia triflora, Syringa emodi,
Rhododendron lowndesii, and Berberis species.

ABIES SPECTABILIS FOREST

Champion: Eastern oak–fir forest, p. 249. Alpine fir–birch forest, p. 269.

Champion's description of alpine fir–birch forest fits well much of the high-
altitude fir forest found in the West Midlands and in the Humla–Jumla area.
His description of eastern oak–fir forest fits fairly well some of the high-altitude
fir forest occurring in the wetter parts of the East Midlands, though I do not

understand why the word 'oak' is included in the title when neither in the examples given by Champion from Sikkim nor in this type of forest in Nepal is any oak present.

Neither description covers the extensive fir forests found at higher altitudes in the more central parts of Nepal. I prefer to group under one heading all the forest in which the fir occurs as a dominant species.

Abies spectabilis, which has also been known as *Abies webbiana*, is found throughout the Himalaya from Afghanistan to Bhutan. In Nepal it is widespread between 10,000 ft and the treeline, and sometimes descends as low as 9,000 ft.

Botanical opinion is not unanimous on the question whether *Abies spectabilis* is specifically distinct from *Abies densa* and *Abies pindrow*, but I understand that at the present time they are generally considered to be three distinct species. Certainly *Abies pindrow* is readily distinguishable when seen in the field; in Nepal it occurs between 7–10,500 ft only in western parts of the country. I deal separately with this low altitude fir under the heading *Abies pindrow* forest.

Abies densa on the other hand is not nearly so distinct from *Abies spectabilis* in habitat or in appearance. In Sikkim both species apparently occur in the upper forest at similar altitudes. It seems possible that some of the fir occurring in the easternmost parts of Nepal is in fact *Abies densa*, but further investigation is needed as to the extent to which it occurs here and the limit of its range westwards. In the following notes I make no further reference to *Abies densa*, but refer to all the high-altitude fir as *Abies spectabilis*.

Abies spectabilis forest is most extensive in the Central Midlands of Nepal. Here between 10–11,500 ft on the south side of the main ranges there is usually a continuous belt of almost pure *Abies spectabilis*. Above 11,500 ft the *Abies* often is superseded by *Betula utilis*, but in some places it ascends to the treeline. Below 10,000 ft it usually gives way to *Tsuga* forest or to *Acer*, *Osmanthus*, and *Magnolia* of the upper temperate mixed broadleaved forest.

This fir forest normally has a dense understory of rhododendrons, and when seen in the spring it is one of the most beautiful sights in Nepal. The upper canopy of the forest is composed almost exclusively of the fir, and the straight-stemmed trees attain a height of 80–100 ft. Beneath their dark green foliage glow the deep crimson flowers of *Rhododendron barbatum*, and the forest floor is brightened by drifts of *Primula petiolaris*, *Primula gracilipes*, and *Primula denticulata*. In early April the forest is scented with the fragrance of the flowers of *Daphne bholua*.

In the Central Midlands the rhododendrons occuring in *Abies* forest are limited to *Rhododendron barbatum, R. campanulatum, R. arboreum*, and in a few places the Nepalese endemic *R. cowanianum*. In the East Midlands *R. hodgsonii, R. grande*, and *R. falconeri* also occur in the forest, and these latter species in some places form dense tangled thickets 30 ft high.

Broadleaved trees are not common in this *Abies–Rhododendron* forest and mostly are confined to clearings. The ones most frequently found are *Betula utilis* and species of *Sorbus* and *Acer*. In contrast to *Abies* forest in the West Midlands and the Humla–Jumla area, *Quercus semecarpifolia* is not a normal component of the *Abies–Rhododendron* forest in the Central and Eastern Midlands. Trees of *Juniperus recurva* on the other hand are quite common, and where *Abies* has been cleared one frequently finds secondary forest consisting of almost pure juniper, particularly on slopes which face south and east.

In *Abies–Rhododendron* forest in the Central and East Midlands I have noted the following species:

1 Abies spectabilis, Tsuga dumosa.

2 Betula utilis, Juniperus recurva, Sorbus cuspidata, Sorbus foliolosa, and species of Acer.

3 Rhododendron arboreum, R. barbatum, R. campanulatum, R. grande, R. falconeri, R. hodgsonii, R. cowanianum, Sorbus foliolosa, Litsea sericea, Viburnum cordifolium, Piptanthus nepalensis, Daphne bholua, Lonicera angustifolia, and species of Arundinaria.

4 Clematis montana.

Where the *Abies* is burnt, dense thickets of bamboo often spring up. An all too common sight at these altitudes is a hillside on which stand the dead and fire-blackened trunks of conifers with a dense canebrake beneath them. Often the forest has been destroyed deliberately, but the cane prevents any use of the slopes for grazing. Even in living conifer forest with a continuous canopy there are many places where canebrakes are so dense as to prevent all movement off the tracks.

Abies forest is less widespread in the East than in the Central Midlands. In areas of very heavy rainfall on the upper Arun and Tamur this type of forest to a large extent is replaced by *Rhododendron* forest (see the section dealing with *Rhododendron* forest, p. 100), in which *Abies* occurs only singly or in small groups scattered amongst the dense *Rhododendron* shrubberies. There are intermediate stages between *Abies* forest and *Rhododendron* forest, but in their typical forms they are very distinct.

Pure *Abies* forest is also much less extensive in the West Midlands and in the Humla–Jumla area, for here it is very often broken up by stands of *Quercus semecarpifolia* which on sunny south-facing slopes may ascend to the treeline, and by *Betula* forest which on unstable slopes and in gulleys where the snow lies late penetrates deep down into the *Abies* zone. Champion refers to this mixture of birch, fir, and oak as 'Alpine fir–birch forest'.

In many places the three species separate out into pure stands, but they also very frequently occur mixed. In the mixed forest the upper canopy is composed largely of fir, with the oak forming the understory, but sometimes the oak and even the birch reach the upper canopy. The almost constant presence of oak either in the *Abies* forest itself or on adjacent slopes distinguishes this type of forest from the typical *Abies* forest of the wetter parts of Nepal. In *Abies* forest in the Jumla area I have noted the following species:

1 Abies spectabilis.

2 Quercus semecarpifolia, Betula utilis, Acer caesium, Acer pectinatum, Acer acuminatum, Prunus cornuta, Prunus rufa, Sorbus foliolosa, Sorbus cuspidata, Rhododendron arboreum. Species of Taxus particularly abundant on limestone.

3 Rosa macrophylla, Rosa sericea, Deutzia corymbosa, Ribes emodense, Ribes glaciale, Lonicera webbiana, Lonicera myrtillus, Lonicera quinquecocularis, Lonicera lanceolata, Lonicera hispida, Cotoneaster acuminata, Piptanthus nepalensis, Euonymus porphyreus, Viburnum cotinifolium, Daphne bholua, Rhododendron campanulatum, Rhododendron barbatum, Spiraea bella, Leycesteria formosa, and species of Salix, Rubus, Philadephus, Berberis, and Arundinaria. The latter is very dense in places.

4 Clematis montana, Clematis barbellata, Schizandra grandiflora.

In some places in the West Midlands rhododendrons in the *Abies* forest remain almost as abundant as they are in more eastern parts of the country, even though the number of species is much reduced. In the Humla–Jumla area, however, they are much less plentiful, and may sometimes be entirely absent. For example, on a north slope near the Rara lake *Abies* forest occurs between 10–11,500 ft in which there is neither oak nor rhododendron. Young seedling firs are the only trees at all common in the understory, and the moss-covered ground beneath the tall straight-stemmed trees is very open. I noted here the following species:

1 Abies spectabilis.

2 Betula utilis, Taxus species.

3 Sorbus foliolosa, Rosa macrophylla, Daphne bholua, Lonicera myrtillus, Arundinaria species.

Abies is found in decreasing quantity as one approaches the heads of the dry inner valleys of the Western half of Nepal. Here on the fringe of the steppe country it is largely replaced in the upper forest by *Pinus excelsa*, *Betula utilis*, and *Juniperus wallichiana*. Its absence may be due to burning as well as to reduced rainfall.

TSUGA DUMOSA FOREST

Champion: Eastern oak–hemlock forest, p. 246.

Tsuga dumosa, the Himalayan hemlock, ranges from Kumaon eastwards through Nepal, Sikkim, and Bhutan to the Mishmi hills and upper Burma. The most westerly locality from which it has been recorded is in the Mahakali valley just to the west of the Nepal–Kumaon border. It does not occur in the NW. Himalaya.

In Nepal *Tsuga dumosa* is widespread at altitudes between 7–11,000 ft.

The examples of *Tsuga* forest given by Champion are from the very wet outer hills of the Darjeeling district. They are not at all typical of most of the *Tsuga* forest found in Nepal.

I have seen *Tsuga* occurring with *Lithocarpus pachyphylla* to form forest of Champion's 'Eastern oak–hemlock' type only in the extreme east of the country on the ridges north of Ilam. Here between 8–9,500 ft *Lithocarpus pachyphylla* occurs in some quantity, together with *Quercus lineata* and other East Himalayan broadleaved species such as *Michelia doltsopa*, *Rhododendron grande*, and *R. falconeri*. In parts of this forest there are stands of tall *Tsuga* trees.

Elsewhere in the East and Central Midlands the *Tsuga* sometimes mixes with *Quercus lamellosa*, but it occurs much more typically mixed with the *Acer*, *Magnolia*, and *Osmanthus* of the upper temperate mixed broadleaved forest. The broadleaved trees predominate in the damper hollows, while the *Tsuga* often separates out into pure stands on the ridges and the drier ground. With much *Rhododendron barbatum* and *R. hodgsonii* beneath it the *Tsuga* forms a kind of *Tsuga–Rhododendron* forest very similar to the *Abies–Rhododendron* forest described elsewhere. This forest is very beautiful in the spring when the rhododendrons are in flower.

In the West Midlands *Quercus lamellosa* and most of the East Himalayan species found in the upper temperate broadleaved forest are absent. They are replaced on northern slopes and in damper places by *Quercus dilatata* and *Aesculus–Juglans–Acer* forest, amongst which a few trees of *Tsuga* can be found. *Tsuga*, however, also forms much almost pure forest here, and this pure forest is much more extensive in the West Midlands and in some of the eastern inner valleys than it is on the wet southern slopes of the main ranges in the eastern parts of the country. Under very wet conditions the *Tsuga* appears to find difficulty in competing successfully with the broadleaved species.

Tsuga forest can be found even in the extreme west of the country. Quite extensive *Tsuga* forest occurs in the Marma district close to the most westerly locality from which this species has been recorded.

This western form of *Tsuga* forest usually meets with *Abies* and *Betula* forest at about 10,500 ft. Its lower part often meets with *Aesculus–Juglans–Acer* forest at about 8,000 ft. *Quercus semecarpifolia* is usually present in the vicinity on all the drier south-facing slopes. The *Tsuga* trees attain a height of 80–100 ft. I have noted the following species in this type of forest in the West Midlands.

1 Tsuga dumosa. A few Abies spectabilis, Betula utilis, Quercus semecarpifolia.

2 Acer sterculiaceum, Acer cappadocicum, Acer acuminatum, Sorbus cuspidata, Prunus cornuta, Ilex dipyrena, Symplocos theaefolia, Lindera pulcherrima, Rhododendron arboreum.

3 Viburnum cordifolium, Viburnum stellulatum, Daphne bholua, Piptanthus nepalensis, Rhododendron barbatum, Rosa sericea, Rosa macrophylla, Hydrangea heteromalla, Cornus macrophylla, Jasminum humile, Lonicera webbiana, Ribes glaciale, and species of Sarcococca, Berberis, Rubus, Salix, and Arundinaria.

4 Clematis montana, Clematis barbellata, Hedera nepalensis, Euonymus echinatus, Holboellia latifolia, Schizandra grandiflora.

The composition of this western form of *Tsuga* forest differs from the eastern form in which dense thickets of *Rhododendron* often exclude other shrub species. *Tsuga–Rhododendron* forest of this type does occur in a few places in western parts of the country, but there the *Rhododendron* species are limited to *R. arboreum*, *R. barbatum*, and *R. campanulatum*. In the east *R. hodgsonii*, *R. falconeri*, and *R. grande* can also be found beneath the *Tsuga*.

Although *Tsuga* forest is quite extensive on the southern sides of the chain of lekhs which separate the Humla–Jumla area from the West Midlands, in the Humla–Jumla area itself I have seen *Tsuga* forest of any extent only in one or two of the side valleys of the Humla Karnali. *Tsuga*, however, is not uncommon as a minor component of the *Picea smithiana* and *Abies pindrow* forests which are found elsewhere in the area. In these magnificent forests, where trees of these two dominant conifers often attain 150 ft, it is interesting to note that fully grown *Tsuga* trees are unable to reach the top canopy.

PINUS EXCELSA FOREST

Champion: Lower blue pine forest, p. 260. High level blue pine forest, p. 272. Western mixed coniferous forest, p. 240.

Pinus excelsa Wall. ex D. Don is perhaps more correctly known as *Pinus wallichiana* A. B. Jackson, but I refer to it here by the former name which I think is more widely used.

Pinus excelsa ranges from Afghanistan through the Himalaya to Bhutan and SE. Tibet. K. C. R. Choudhury states[16] that this pine does not occur in Sikkim naturally. In Nepal its lower altitude limit is at about 6,000 ft; its upper limit is at the treeline. I have seen this pine at 14,500 ft in the Sibu khola to the NE. of Kanjiroba Himal, but this altitude is exceptional.

The Midlands

Pinus excelsa is found in most parts of Nepal, but it is very much less abundant in the Midlands than in other parts more sheltered from the full force of the monsoon rains. In the Midlands it is not uncommon in the big river valleys, coming in above *Pinus roxburghii* at about 6,500 ft. In these places the presence of a few oak trees often makes one suspect that the pine has invaded the slopes only after the original oak forest has been destroyed.

The pine is an early coloniser of abandoned fields and grazings, and in the

Midlands much of the ground on which it occurs is of this type. Indeed, I cannot remember ever having seen this pine in the Midlands forming forest in a habitat which I have not suspected to have been in some way modified by the activities of man.

Pine forest here tends to be very open, often with scrub oak surviving amongst it. (See for example the section dealing with *Quercus semecarpifolia* forest, p. 85.)

The Humla–Jumla area

Here *Pinus excelsa* is much more widespread.

Below 10,500 ft much of the forest cover, particularly on south-facing slopes, consists of almost pure pine forest. *Picea smithiana* and *Abies pindrow* are largely confined to north or west faces, and even here the pine is abundant where the forest has been damaged.

Pinus excelsa, unlike mature trees of *Pinus roxburghii*, is not immune to fire, but its ability quickly to recolonise open slopes gives it an advantage over other conifer species.

In view of the secondary origin of much of the pine forest it is not surprising that there is a tendency in many places for all the trees to be of the same age and height. Near the Rara lake, for example, there are some dense thickets of pine about 30 ft tall which very much resemble a plantation. The lower branches of the closely packed trees are dead and lichen-covered, and there are no associated trees or shrubs except a few *Juniperus wallichiana*.

When they mature these even-aged pine forests produce fine stands of trees over 100 ft tall. There are plenty of good examples of mature forest in the Humla–Jumla area, but in the Midlands they are much less common, probably because the demand for timber is more acute here.

As well as forming almost pure pine forest *Pinus excelsa* in the Humla–Jumla area often occurs mixed with *Picea smithiana* or *Abies pindrow* to form what Champion refers to as western mixed coniferous forest. In the more restricted localities where *Cedrus deodara* and *Cupressus torulosa* occur the pine can also be found mixed with them. Above 10,000 ft *Abies spectabilis* becomes increasing predominant, and at about 11,000 ft the pine usually disappears.

These mixed forests of the Humla–Jumla area are very attractive, and are interspersed with meadows bright with summer flowers. In listing the species which occur in these forests I have included a number which in fact are more closely associated with the clearings which so often intrude into them. The pine itself is to a large extent associated with these places. (For descriptions of undisturbed conifer forest in this area see the sections dealing with *Picea smithiana* forest, p. 115, and *Abies pindrow* forest, p. 117.)

In the Humla–Jumla area between 7–10,500 ft in this mixed and broken type of forest in which *Pinus excelsa* is very widespread and locally dominant I have noted the following species:

1 Pinus excelsa, Picea smithiana, Abies spectabilis, Abies pindrow, Cedrus deodara.

2 Quercus semecarpifolia, Betula utilis, Alnus nepalensis, Corylus colurna, Sorbus cuspidata, Sorbus lanata, Sorbus foliolosa, Aesculus indica, Juglans regia, Juniperus wallichiana, Acer caesium, Acer cappadocicum, Acer pectinatum, Prunus cornuta, Ulmus wallichiana, Symplocos crataegoides, Euonymus tingens, Euonymus fimbriatus, Rhododendron arboreum, Cornus capitata, Salix species.

3 Buddleja tibetica, Rosa macrophylla, Rosa brunonii, Colquhounia coccinea, Caragana brevispina, Daphne bholua, Lonicera myrtillus, Lonicera quinquelocularis, Viburnum cotinifolium, Viburnum grandiflorum, Jasminum humile, Rhododendron campanulatum, Rhododendron anthopogon, Rhododendron lepidotum, Lyonia villosa, Acanthopanax cissifolius, Ribes glaciale, Piptanthus nepalensis, Leycesteria formosa, Spiraea canescens, Prinsepia utilis, Hydrangea heteromalla, Rhus cotinus, Syringa emodi, and species of Xanthoxylum, Elaeagnus, Berberis, Deutzia, Indigofera, Leptodermis, Rubus, Skimmia and Arundinaria.

Inner valleys, and edges of the arid zone

In most of the inner valleys of the eastern half of Nepal the significant reduction in rainfall which takes place towards the valley heads occurs at an altitude higher than the treeline, so that the forest found here differs little from that found at similar altitudes in the Midlands. Where *Pinus excelsa* occurs in these places it is mostly of the medium-altitude river-valley type which I have described for the Midlands.

In the inner valleys of the western half of the country the position is different, for the diminution in the rainfall begins at lower altitudes.

As one ascends any of the valleys leading up towards the dry areas which lie along the border in West Nepal one can see the forest flora gradually becoming more limited, until ultimately of the forest trees only *Pinus excelsa, Cupressus torulosa, Betula utilis, Juniperus wallichiana,* and a few *Abies spectabilis* survive.

The absence of other tree species must be due in part to the dry hard conditions prevailing here. The dryness of the undergrowth, however, favours the spread of fire, and I suspect that the predominance of the pine in many of these places is due more to fire than to climatic causes. In particular, the relative scarcity of *Abies spectabilis* at high altitude in these dry places appears to result largely from fire.

I have seen this transition in most of the valleys which lead up towards Mustang and Dolpo; Mugu khola, Langu khola, Phoksumdo khola, Tarap khola, Barbung khola, and the Tukucha valley; and I assume that the same transition takes place at the head of the Marsyandi valley. Pine and birch compose the main part of such forest as survives in these valleys. The cypress does not go much above 10,500 ft and the juniper is mostly limited to stunted trees scattered individually across the stony hillsides.

Under these dry conditions the pine ascends to very high altitudes, and as already mentioned I have seen it as high as 14,500 ft. More often, however, the pine drops out at about 12,000 ft, leaving thickets of birch above. In *Betula–Pinus* forest in the Mugu khola between 10,500–12,000 ft I noted the following species.

1 Pinus excelsa, Betula utilis. A very few Abies spectabilis, Picea smithiana.

2 Rosa sericea, Rosa macrophylla, Lonicera myrtillus, Lonicera obovata, Lonicera webbiana, Viburnum cotinifolium, Syringa emodi, Spiraea bella, Spiraea arcuata, Cotoneaster acuminata, Cotoneaster microphylla, Potentilla fruticosa, Plectranthus pharicus, Rhododendron lepidotum, Juniperus wallichiana, Juniperus communis, and species of Leptodermis, Salix, and Berberis.

At somewhat lower altitudes in this khola there is a good deal of almost pure pine forest. In places here there is also some *Cupressus torulosa*, and I suspect that the cypress has to a large extent been displaced by the pine as a result of fire. This pine forest is very open with trees not much exceeding 50 ft in height. Here between 9,500–10,000 ft I noted the following species:

1 Pinus excelsa.

2 Rosa sericea, Plectranthus pharicus, Viburnum cotinifolium, Wikstroemia canescens, Syringa emodi, Spiraea canescens, Lonicera quinquelocularis, Ribes alpestre, Juniperus wallichiana, and species of Salix, Berberis and Leptodermis.

PICEA SMITHIANA FOREST

Champion: Western mixed coniferous forest, p. 240.

In Nepal the individual conifer species which compose Champion's forest type very commonly form a mixed forest in the manner he describes for the Western Himalaya, but they do also occur in pure stands over limited areas. I find it easier to describe the characteristics of these various conifers by devoting a separate section to each species.

Picea smithiana (Wall.) Boissier, the West Himalayan spruce, is also known as *Picea morinda* Link. The species ranges from Afghanistan through the NW. Himalaya to Nepal.

Picea spinulosa, the East Himalayan spruce, ranges from Sikkim eastwards through the Eastern Himalaya. It has not been recorded from Nepal.

In Nepal *Picea smithiana* is found only in those parts of the country which are to some extent sheltered from the full force of the monsoon rains. It is abundant in the Humla–Jumla area, penetrating up the Humla Karnali to Simikot and up the Mugu Karnali almost to Mugu; it ascends the upper Bheri for some distance beyond Tarakot; it occurs in the side valleys of the upper Kali Gandaki north of Tukucha; and J. Kawakita reports[17] that it grows in the upper valleys both of the Marsyandi and of the Buri Gandaki. The latter locality appears to be the most easterly place from which this species has been recorded.

I have never seen *Picea smithiana* anywhere in the Midland areas. One of the major vegetational contrasts between the south and the north sides both of the Dhaulagiri–Annapurna massif and of the lekhs which lie to the south of the Jumla area is the abrupt disappearance of the spruce on the south sides.

I confine my comments to the *Picea* forests of the Humla–Jumla area. Here the spruce ranges in altitude between 7,500–11,000 ft. Frequently it occurs together with *Pinus excelsa* to form the very mixed type of forest which Champion refers to as western mixed coniferous forest and which I further describe in the section dealing with *Pinus excelsa*. Cultivation in the Humla–Jumla area takes place at altitudes up to 10,000 ft, and in consequence, the original forest cover has been much modified. This modification works to the benefit of the pine and the detriment of the spruce which is now largely confined to undisturbed north faces. On these faces the pine is often absent, or limited to the outer fringes of the spruce forest.

Seen from a distance these spruce forests give one the impression that they are composed of this species alone, for frequently they contain magnificent trees 150 ft in height and the high spruce canopy hides all else. Closer inspection, however, reveals a great many other tree species, both in the glades and gulleys and as an understory beneath the spruce. *Quercus dilatata, Quercus semecarpifolia, Juglans regia*, and *Tsuga dumosa* may all attain here a height of 100 ft and yet easily be overtopped by the spruce. This type of forest in which spruce predominates usually passes into *Abies spectabilis* and *Betula utilis* above 10,500 ft.

In *Picea* forest between 9–10,000 ft on a north slope near the Rara lake I recorded the following species:

1 Picea smithiana. A few Pinus excelsa. Abies spectabilis in the upper part.

2 Quercus semecarpifolia, Betula utilis, Populus ciliata, Juglans regia, Prunus cornuta, Sorbus cuspidata, Acer caesium, Acer pectinatum, Taxus species.

3 Sorbus foliolosa, Viburnum cotinifolium, Caragana brevispina, Jasminum humile, Ribes glaciale, Euonymus porphyreus, and species of Berberis, Salix, and Arundinaria.

Pure spruce forest appears to be rather more extensive in Humla than in Jumla, and there are some very fine stands in the side valleys which lead into the Humla Karnali. *Abies pindrow* also occurs here, and in places the two species combine to form what is without doubt the finest coniferous forest in Nepal. Sometimes these 150-ft conifers are closely packed, with little beneath them except a carpet of fallen needles; more often the big stands are broken by open glades with a more diverse flora. (For a further description of this mixed forest see the section dealing with *Abies pindrow* forest, on p. 117.)

On a slope in Humla between 8,500–10,000 ft where *Picea smithiana* predominated but *Abies pindrow* was also present, I noted the following species:

1 Picea smithiana, Abies pindrow. Pinus excelsa at the edge of clearings.

2 Tsuga dumosa, Populus ciliata, Aesculus indica, Juglans regia, Quercus semecarpifolia, Acer cappadocicum, Acer acuminatum, Betula utilis, Sorbus cuspidata, Prunus cornuta, Taxus species.

3 Corylus colurna, Symplocos crataegoides, Cornus macrophylla, Lyonia ovalifolia, Rhododendron arboreum, Meliosma dilleniifolia, Rosa sericea, Rosa macrophylla, Deutzia hookeriana, Syringa emodi, Hydrangea heteromalla, Lonicera quinquelocularis, Lonicera webbiana, Cotoneaster acuminata, Sorbus foliolosa, Caragana brevispina, Viburnum stellulatum, and species of Indigofera, Arundinaria, Ribes and Skimmia.

4 Clematis barbellata, Schizandra grandiflora, Pentapanax leschenaultii, and species of Vitis.

Picea smithiana ascends a long way up some of the dry river valleys which lead from the Humla–Jumla area towards the Tibetan frontier. The spruce often consists only of a few trees scattered amongst the *Pinus excelsa* which predominates in many of these places (see the section dealing with *Pinus excelsa* forest, on p. 111). Occasionally, however, there are small stands of almost pure *Picea*. One such stand occurs at 10,000 ft near Simikot, where open and almost pure spruce forest covers an area of about 4 acres. The trees here are not more than 70 ft tall, and conditions seem too severe for the survival of the broadleaved species of tree which are the normal components of moist spruce forest, such as *Quercus semecarpifolia, Juglans regia,* or *Aesculus indica.* I noted here the following species:

1 Picea smithiana, a few Pinus excelsa.

2 Juniperus wallichiana, Taxus species, Cotoneaster frigida, Cotoneaster acuminata, Wikstroemia canescens, Viburnum cotinifolium, Colquhounia coccinea, Rosa macrophylla and species of Indigofera, Berberis and Deutzia.

ABIES PINDROW FOREST

Champion: Western oak-fir forest, p. 244. Western mixed coniferous forest, p. 240.

In the first of these two forest types of Champion's the predominant species are *Abies pindrow* and *Quercus semecarpifolia.* In the forests of West Nepal where *Abies pindrow* occurs *Quercus semecarpifolia* is quite commonly present, but usually only as one of a number of other broadleaved species. The comments I make concerning the second of Champion's above-mentioned forest types in the section dealing with *Picea smithiana* also apply in this section.

There appears to be some doubt as to whether *Abies pindrow* is or is not specifically separable from *Abies spectabilis.* As seen in the field in Nepal, however, the two trees do differ very much both in appearance and in habitat.

Abies spectabilis, the high-altitude fir, is common in the upper forest in most parts of Nepal, and ascends to the treeline. It is not usually found below 10,000 ft, but where occasional trees occur rather below that altitude they retain the appearance characteristic of the tree at higher altitudes.

Abies pindrow, the low-altitude fir, in Nepal is confined to north- and west-

facing slopes in a few areas in the western parts of the country. It occurs between 7–10,000 ft, and does not reach the treeline. It is a very tall tree, often
attaining 150 ft, and because its branches are borne close in to the trunk it has a
columnar appearance. In contrast to this, *Abies spectabilis* does not usually much
exceed 80 ft in height, its branches are widely spreading, and its leaves are much
more stiff and rigid than the rather lax leaves of *Abies pindrow*.

Whether these apparent differences are due to specific distinction or to
variation in the growing conditions at the altitudes at which the trees occur is an
interesting question for a taxonomist. In the field in Nepal they differ so much
in appearance and habitat that obviously it is convenient to refer to the two trees
by different names.

West of the Karnali *Abies pindrow* grows between 7–9,500 ft, always on north
or west faces and usually in gulleys where the soil is damp. In most places here
it occurs only as a widely-spaced component in an otherwise exclusively broadleaved *Aesculus–Juglans–Acer* forest. The fir looks very incongruous, its tall
columns towering at intervals over the broadleaved canopy beneath.

At similar altitudes in Humla *Abies pindrow* is a good deal more abundant,
though in the Jumla area I have seen it only rarely. In Humla the upper canopy
of the forest in some places is formed exclusively of the fir, though more
commonly it is mixed with *Picea smithiana* in roughly equal proportions. Both
trees attain a height of 150 ft, and both have a columnar appearance. Standing
in the forest and running one's eye upwards along the great soaring trunks one
experiences a visual pleasure similar to that which can be felt amongst the piers
and vaulting of a Gothic cathedral.

Spruce and fir in places grow closely packed together, and here the forest
floor beneath them is dark and shrubless. More often there are a number of
broadleaved trees beneath the coniferous upper story, and in some places the
growth of the undershrubs is dense. *Vitis divaricata* is prominent in this forest,
often festooning the big conifers to a height of 80 ft.

In Humla *Abies pindrow* forest overlaps with and is not clearly distinguishable
from *Picea smithiana* forest. On slopes where *Abies pindrow* was generally predominant but *Picea smithiana* also numerous I noted the following species:

1 Abies pindrow, Picea smithiana, a few Pinus excelsa.

2 Tsuga dumosa, Quercus dilatata, Quercus semecarpifolia, Aesculus indica,
Juglans regia, Populus ciliata, Acer cappadocicum, Acer acuminatum, Betula
utilis, Prunus cornuta, Sorbus cuspidata, Corylus colurna, Taxus species.

3 Cornus macrophylla, Lyonia ovalifolia, Rhododendron arboreum, Symplocos crataegoides, Meliosma dilleniifolia, Rosa macrophylla, Deutzia hookeriana, Viburnum stellulatum, Viburnum cotinifolium, Staphylea emodi, Philadelphus tomentosus, and species of Ribes and Indigofera.

4 Hedera nepalensis, Schizandra grandiflora, Vitis divaricata.

CEDRUS DEODARA FOREST

Champion: Moist deodar forests, p. 237.

Cedrus deodara ranges from Afghanistan through the NW. Himalaya to West Nepal. In the Humla–Jumla area it is common as a village tree, but occurs naturally only in a few places. The principal localities in which I have seen the cedar occurring naturally here are as follows:

The valley of the Tila khola
SW. of Jumla, 6,500–8,000 ft.

For a distance of about 15 miles cedar forest grows intermittently on the valley sides leading down to the river. Much of the rock here is limestone, and the valley is windy. The slopes are dry and to a large extent deforested.

In some places the forest consists almost purely of cedar, in others there is a considerable admixture of *Pinus excelsa*. Burning appears to be frequent, and the cedar forest is seen at its best on steep rocky slopes which escape the fire. Much of the forest is very open, and often it alternates with grassland. Shrubs are few, but the list below of those occurring here is by no means complete, because I visited this locality in March at a time when most of the shrubs were leafless. I noted the following species:

1 Cedrus deodara, Pinus excelsa.

2 Rosa sericea, and species of Salix, Berberis, Indigofera, Spiraea, Prunus, Plectranthus, Lonicera, and Cotoneaster.

3 Clematis montana.

The valley of the upper Bheri
Between Tibrikot and Tarakot, 7,500–9,500 ft.

This is the most easterly locality from which cedar forest has been recorded. The valley here is very windy, and the treeless lower slopes running down to the river are covered only with a dry steppe–like association of *Artemisia* and grasses. Above this there is a belt of cedar extending intermittently along the valley for about 15 miles.

In most places the cedar belt is only a few hundred feet wide, being succeeded above by *Pinus excelsa* or *Picea smithiana*. The open forest is composed almost exclusively of small cedar trees.

The cedar penetrates a short distance up some of the side valleys, and in the rather wetter conditions prevailing in these places mixes with *Pinus* and *Picea* and attains a height of 100 ft. But both here and in the Tila khola the cedar for the most part is confined to the dry windy main valley.

The valley of the lower Bheri

North of Gurta, 8,000 ft.

Rainfall here is higher. The cedar trees occur amongst *Pinus, Picea, Quercus* and other broadleaved trees in some of the side valleys.

The Humla Karnali valley

Ten miles below Simikot, 7–8,000 ft.

I have seen through binoculars but not confirmed on the ground the presence here of some stands of cedar. They are mostly on steep rocky slopes, with *Pinus excelsa* above and *Pinus roxburghii* below.

CUPRESSUS TORULOSA FOREST

Champion: Cypress forest, p. 256.

Cupressus torulosa ranges from Chamba in the NW. Himalaya to Nepal. Formerly some of the specimens of cypress collected from SE. Tibet and Western China were named as *Cupressus torulosa*, but it seems that this is incorrect and that they are all either *Cupressus fallax* or *Cupressus duclouxiana*.[18] I know of no record of *Cupressus torulosa* occurring in Nepal in natural forest anywhere east of Tukucha, and on this distribution one would suppose that the species is limited to the Western Himalaya, for it appears to be unknown from Sikkim and Bhutan. Rather surprisingly, however, N. L. Bor reports cypress trees growing on limestone in the Aka hills of the Assam Himalaya,[19] and states that the species is *Cupressus torulosa*. If these trees are indeed *Cupressus torulosa* the species appears to have an unusually broken distribution.

In Nepal I have seen *Cupressus torulosa* on the south side of the main ranges only in the extreme west of the country, in the parts which lie to the west of the Karnali. Here it is confined to the belt of limestone which runs from the Kumaon border through Marma to the Seti khola.

Elsewhere in Nepal the cypress is not confined to the limestone, but the localities in which I have seen it growing are all to some extent sheltered from the full force of the monsoon rains. It is quite common as a village tree in the Humla–Jumla area, but it is not a normal component of the mixed conifer forests of this district. It does, however, form apparently natural forest in one small area in the Khater khola which leads from the Rara lake to the Karnali. North of the Dhaulagiri massif it occurs in the upper Bheri valley, the Suli Gad, and the Tarap khola, and also around Tukucha in the dry windy valley of the Kali Gandaki.

West of the Karnali the cypress is always found on slopes which face south or east. In this limestone country it appears to have suffered much from fire and is now mostly confined to steep rocky faces where it forms very open forest with little combustible undergrowth beneath. In the few localities where it occurs on more gradual slopes the cypress forest has a richer composition rather similar to that described below for the Suli Gad.

Elsewhere in Nepal the cypress is not confined to south or east faces, but fire often confines it to riverside banks and steep rocks. Sometimes stands of cypress grow in isolation in fireproof places surrounded by treeless grass slopes; sometimes the intervening slopes have been recolonised by *Pinus excelsa*.

Many of the places in which it occurs are close to the treeless Tibetan steppe country. The demand for timber and firewood is acute here and the cypress has suffered much from felling and lopping. The valley floor from Tukucha northwards towards Jomsom is covered with the remains of cypress forest, the battered stumps of which in many cases are only a few feet high. *Juniperus wallichiana* also occurs here, and from a distance the mutilated trees of the two species are not easily distinguished.

In most places the cypress occurs at altitudes between 7–9,500 ft. Occasionally it ascends rather higher, and at Kakotgaon in the Barbung khola there is a small and stunted cypress forest on the edge of the steppe country at a height of about 11,000 ft. In general, however, cypress tends to be replaced by juniper at altitudes above 10,500 ft.

The largest extent of undamaged cypress forest I have seen is in the Suli Gad which leads from the upper Bheri to the Phoksumdo lake. Some of the cypress here is confined to steep rocky slopes with the intervening ground covered with either grass or *Pinus excelsa* in the manner described above, but on slopes facing north and west, and hence presumably rather damper and less subject to the spread of fire, there is extensive cypress forest with a much richer composition. The trees are about 70 ft tall and the dark green of their foliage contrasts with the vivid blue-green of the waters which tumble down from the lake. Set between high brown cliffs and clothed with fine cypress forest this valley is one of the most beautiful in Nepal.

In the forest here I have noted the following species:

1 Cupressus torulosa.

2 Wikstroemia canescens, Colquhounia coccinea, Spiraea sorbifolia, Rosa macrophylla, Abelia triflora, Syringa emodi, Viburnum cotinifolium, Plectranthus pharicus, and species of Berberis, Cotoneaster, Deutzia, Lonicera, Rubus, Indigofera, and Arundinaria.

3 Cynanchum auriculatum, Clematis phlebantha.

LARIX FOREST

Champion: Larch, p. 273.

Larch occurs in the Eastern Himalaya from Nepal eastwards through Sikkim and Bhutan into SE. Tibet and China. It is absent from the NW. Himalaya.

In the extreme east of Nepal close to the Sikkim border larch occurs in the Kambachen valley and also in the nearby Simbua khola. In Central Nepal it has been recorded in the Langtang valley, near Rasua Garhi, and in the Shiar khola; and J. Kawakita reports it as growing in the upper Buri Gandaki valley.[20] In the gap which intervenes between Kambachen and Langtang it has been reported on the Tibetan side of the frontier in the Rongsha Chu.[21]

The larch of Sikkim and Bhutan is *Larix griffithiana*, and a gathering made at Kambachen in Nepal is undoubtedly of this species. It is therefore somewhat surprising that gatherings made from Langtang, Rasua Garhi, and Shiar khola have been identified as being *Larix potaninii*. This is a western Chinese species which appears to be unrecorded anywhere else in the Himalaya.

Our present incomplete knowledge of the flora of the Eastern Himalaya certainly leaves a very real possibility that although it has not as yet been recorded there, *Larix potaninii* does in fact occur elsewhere in the country which lies between Nepal and China. At the same time the taxonomic distinctions between the two species do not seem to be so great as to preclude errors in identification. I am not qualified to venture an opinion on the accuracy of the identifications, but I must point out that if this larch which occurs in Central Nepal is indeed *Larix potaninii* its pattern of distribution is very unusual.

I have seen the larch both at Kambachen and Langtang. Nowhere in the Langtang does the larch form pure forest. A few trees can be found as low as 9,500 ft, growing singly on ground which has mostly been cleared for grazing, and attaining a height of 70 ft. On the south side of the valley much of the forest remains uncut, consisting principally of *Tsuga dumosa* below 10,000 ft and *Abies spectabilis* above this height. Larch grows scattered throughout this forest, and ascends to 12,250 ft at Kyansin. At this height the trees are only 15–30 ft tall, growing on the moraine with shrubs of *Juniperus wallichiana* and *Juniperus squamata* and a few trees of *Betula utilis*.

From Kawakita's account it seems that larch forest in the Shiar khola is more extensive. He reports 'a large aggregation of larch forest above Cchokang, *ca.* 3,150 m.'[22]

My comments on the larch which occurs in the Kambachen valley are based not on field notes but only on the recollections of fifteen years ago. The larch is quite abundant above 10,000 ft, occurring both with *Abies spectabilis* and *Betula utilis*, and also in almost pure stands. Below the village of Ghunsa the trees are 70 ft tall, and in July the ground beneath them is bright with the pink flowers of *Notholirion macrophyllum*. Beyond the village the larch continues up the valley to a height of almost 13,000 ft, where together with the birch it forms the uppermost forest.

ALNUS WOODS

Champion: Alder woods, p. 259.

Alnus nepalensis ranges from Chamba eastwards through the Himalaya to SE. Tibet and Western China. It is widespread in Nepal between 3–9,000 ft along streams and in places where there is permanent water. Exceptionally I have seen this species in the outer foothills as low as 1,500 ft. Trees can attain a height of 100 ft, and often are unbranched until shortly below the crown.

The range of *Alnus nitida*, the West Himalayan alder, is stated in the *Flora of British India* as being from Kashmir to Kunawar. One gathering from West Nepal has been identified as this species. The locality where it was collected is in the valley of the Mugu Karnali. Here the riverside between 7–8,000 ft is lined with this species of alder. The trees are up to 80 ft tall, and extend intermittently for some miles along the river.

Although I know of no other records of the occurrence of *Alnus nitida* in Nepal, perhaps in the future it may be found in other western parts of the country. It does not differ much from *Alnus nepalensis* and could easily be confused with it in the field. There is no doubt, however, that the common alder of Nepal is *Alnus nepalensis*, and it is to this species alone that I refer in the notes below.

Much of the country at the altitudes at which *Alnus nepalensis* occurs is deforested and under cultivation. In the Central and East Midlands the alder is often present in wet places in *Schima–Castanopsis* forest or in the temperate deciduous forest, but it occurs most prominently in the form of small isolated woods amongst the fields and on unstable ground which is unsuitable for cultivation. Local people value the trees for giving some stability to slopes which in the rains have a tendency to slip and erode. These alder woods are so limited in extent and so heavily grazed by village cattle that they have scarcely any associated shrub flora. In the extreme east of the country cardamon is commonly cultivated beneath the trees.

In the West Midlands *Alnus nepalensis* occurs in *Quercus dilatata* forest and in *Aesculus–Juglans–Acer* forest. The latter type of forest is seen at its best on flat terraces alongside streams, and often is fringed at the water's edge by *Alnus nepalensis, Betula alnoides*, and sometimes *Populus ciliata* as well. In these places the tall trees form a very open forest, and the shrubs beneath are much reduced by grazing. *Prinsepia utilis, Rosa brunonii, Coriaria nepalensis,* and species of *Xanthoxylum* seem best adapted for survival here.

In the Humla–Jumla area both *Betula alnoides* and *Alnus nepalensis* are largely replaced by *Populus ciliata,* and to some extent by *Hippophae salicifolia.* Alder is only rarely present in wet places in the mixed conifer forests of the area.

POPULUS CILIATA WOODS

Champion: No separate mention.

Populus ciliata ranges from Kashmir through the Himalaya to Bhutan and SE. Tibet.

In Nepal this poplar is confined to the drier areas where it occurs along streams at altitudes mostly between 7–10,500 ft, though at Muktinath it can be found planted along irrigation channels at over 12,000 ft. It is common in the Humla–Jumla area, and it occurs in the inner valleys of the country from the Trisuli river westwards.

This poplar appears to be absent from the wetter parts of the country. I cannot remember ever having seen it in the Central or East Midlands, nor in the inner valleys of the eastern half of Nepal. In these parts another species of poplar occurs occasionally, *Populus glauca,* but it is not common.

Populus ciliata is a common component of the *Aesculus–Juglans–Acer* forest found in the West Midlands and the Humla–Jumla area. This forest occurs on riverside terraces and often is fringed by poplars on the river bank. Around Jumla the poplar frequently attains dominance over an acre or two of riverside flats. The trees here are up to 80 ft in height, and with their corrugated grey trunks and leaves trembling in the wind they are a prominent feature of the district.

The poplar also commonly occurs in the mixed conifer forests of the Humla–Jumla area. In *Picea smithiana* or *Abies pindrow* forest it occurs in the understory of the forest; in the drier *Pinus excelsa* forest it is usually found in the small strips of broadleaved forest which fill the intervening gulleys. In this area the poplar largely replaces *Alnus nepalensis* in all wet places, presumably because it is better adapted to withstand the hard winter conditions.

The Mugu branch of the Karnali provides interesting examples of various

types of riverain forest at increasing altitudes and in increasingly dry conditions. Between 7–8,000 ft alder dominates the river banks; between 8–10,500
ft poplar dominates; beyond this height the poplar is replaced by willows and
Myricaria species which are typical streamside shrubs of the steppe country. In
some places here the poplar forms an acre or two of pure woodland of an open
nature, with very little undergrowth beneath the 60-ft trees. In other places a
few conifers occur in the woodland, and the shrub composition is richer. I
noted here the following species:

1 Populus ciliata. A few Cupressus torulosa, Picea smithiana, Pinus excelsa.

2 Hippophae salicifolia, Myricaria species, Salix species, Rosa sericea, Rosa
macrophylla, Abelia triflora, Wikstroemia canescens, Lonicera quinquelocularis,
Spiraea sorbifolia, Buxus sempervirens, Buddleja tibetica, Colquhounia coccinea
and species of Berberis, Ficus and Indigofera.

3 Jasminum officinale, Clematis grata, Cynanchum auriculatum.

Hippophae salicifolia is a very constant associate of the poplar on riverside
gravel, and buckthorn thickets may be succeeded by poplar woods when the
soil is sufficiently stabilized. (See the next section, dealing with *Hippophae*
scrub.)

HIPPOPHAE SCRUB

Champion: Temperate Hippophae scrub, p. 260.

The *Flora of British India* lists two species of Himalayan buckthorn, *Hippophae
rhamnoides* and *Hippophae salicifolia*. It seems that Hooker was doubtful whether
the latter was specifically distinct, for he wrote 'different as this plant (i.e.
H. salicifolia) looks in its ordinary condition from *Hippophae rhamnoides* I expect
that it will prove a form of that plant due to the moister climate which it
affects'.

In the Natural History Museum there are a number of gatherings of the
genus collected from many parts of the Himalaya, and also from Afghanistan
and Tibet. They have been named by Dr Arne Rousi of Finland. From a study
of these gatherings it is evident that three species are involved.

Hippophae rhamnoides L.

The species is widespread in north temperate zones. Its subspecies *turkestanica* occurs in Afghanistan, the Karakorum, and Baltistan. Its subspecies *yunnanensis* occurs in West China and SE. Tibet. Its subspecies *gyantsensis* occurs at Gyantse in Tibet.

There are no gatherings of this species or its subspecies from Nepal.

Hippophae thibetana Schlecht.

The species is recorded from North China, Tibet, and the Himalaya where it is found from Garhwal to SE. Tibet. There are numerous gatherings from Nepal.

It is common in the dry inner valleys of Nepal between 11–14,500 ft, particularly on old moraines and stony ground. It is a low shrub, usually not more than 2 ft high.

Hippophae salicifolia Don

The species is recorded from Garhwal to Bhutan, and there are several gatherings from Nepal. I have seen it in the West Midlands, both west of the Karnali and around Dhorpatan; in the Humla–Jumla area; at Tukucha; and in the Langtang valley.

The species seems to require a climate which is not too wet. I have never seen it in the Central or East Midlands, nor in the inner valleys at the eastern end of the country.

Hippophae salicifolia occurs both as a low shrub a few feet tall and as a small tree 25 ft tall. The bright orange fruit is retained throughout the winter. It is found between 7–10,500 ft, and often lines the banks of streams. Sometimes it forms almost pure thickets an acre or so in extent on newly-formed alluvial gravel, or on unstable slopes which have water close to the surface. In this respect it resembles *Alnus nepalensis*, a species which it tends to replace in much of the Humla–Jumla area. The buckthorn is here often succeeded by poplar on more stable ground.

In one such riverain thicket at 9,500 ft near Dhorpatan in the West Midlands I noted the following species:

1 Hippophae salicifolia, Populus ciliata.

2 Lonicera myrtillus, and species of Salix, Myricaria, and Berberis.

127

MOIST ALPINE SCRUB

Champion: Moist alpine scrub, p. 273.

Moist alpine scrub occurs above the treeline in all the wetter parts of Nepal. In the inner valleys and the arid zone it is replaced by dry alpine scrub. In the wet areas the alpine scrub ascends only to about 14,500 ft, whereas in Dolpo dry scrub can be found up to 16,500 ft.

The East Midlands

Rhododendron species are always very numerous in the upper forest here, and in some places the top 2,000 ft or so of 'forest' may in fact consist of rhododendron shrubberies with only a few trees of birch or fir scattered amongst them. These dense impenetrable thickets are 5–12 ft high, and cover the slopes in a continuous blanket. Species commonly found here are *Rhododendron campanulatum, R. wallichii, R. campylocarpum, R. wightii, R. fulgens.*

This blanket usually ceases at about 12,500 ft. A few of the *Rhododendron* species mentioned above, and in particular *R. campanulatum*, ascend to 14,000 ft or so, but at this height they are only a minor element in the composition of the vegetation. Alpine herbs predominate over much of the slopes, and on wet grazing grounds species of *Primula* are abundant. In some places there are dense carpets of *Rhododendron anthopogon, R. lepidotum,* and *R. setosum. Juniperus recurva* in dwarf form is common also.

The dividing line between *Rhododendron* forest and moist alpine scrub in many places is not clearcut, and the following list of shrubs I have noted on alpine slopes in the East Midlands contains some which are perhaps more typical of the uppermost *Rhododendron* forest. (See the section dealing with this type of forest, on p. 100.)

> Rhododendron campanulatum, R. wallichii, R. campylocarpum, R. fulgens, R. wightii, R. pumilum, R. setosum, R. anthopogon, R. lepidotum, Juniperus recurva, Salix sikkimensis, Salix calyculata, Lonicera obovata, Lonicera cyanocarpa, Polygonum vacciniifolium, Potentilla fruticosa, and species of Berberis.

Central and West Midlands

Rhododendron species are here a much less prominent feature of the alpine zone. *Rhododendron setosum* has been recorded as far west as the Trisuli river, and a number of the other eastern species can still be found as far west as the Dudh Kosi, but over much of the Central and West Midlands the only three species found at these altitudes are *R. campanulatum*, *R. lepidotum* and *R. anthopogon*.

The uppermost alpine forest here often consists of *Betula utilis*, with shrubberies of *Rhododendron campanulatum* and species of *Sorbus*. Species most typical of the alpine zone here are *Rhododendron lepidotum*, *R. anthopogon*, *Juniperus recurva*, *Lonicera obovata*, *Potentilla fruticosa*, and species of *Salix*.

Humla–Jumla area

This area adjoins the arid alpine steppe country, and in a number of places here the alpine shrubs include some which are more typical of dry alpine scrub. In the wetter parts of the area, however, the alpine shrubs are much the same as in the Central and West Midlands, though rather more limited in species. *Juniperus recurva* is largely replaced by *Juniperus wallichiana*. In the alpine zone of Sisne Himal I noted the following species:

> Rhododendron anthopogon, R. lepidotum, Juniperus wallichiana, Potentilla fruticosa, Lonicera obovata.

DRY ALPINE SCRUB

Champion: Dry alpine scrub, p. 273.

Inner valleys from Langtang eastwards

In the eastern half of Nepal the vegetation of the lower parts of the inner valleys shows a strong contrast to that of the dry upper parts. Below the tree-line the vegetation differs little from that from on the wet southern sides of the main ranges; the alpine flora shows affinities with the flora of Tibet.

The zone of transition usually is between 12–14,500 ft, and is marked by a change in the shrub junipers. Below 12,000 ft *Juniperus recurva* is the normal shrub juniper here; at 12,000 ft in the valley heads *Juniperus wallichiana* begins to appear, and by 14,000 ft it predominates. *Juniperus squamata* also is sometimes present in the zone of overlap between the two other species. Undoubtedly it is rainfall and not altitudinal preference which prevents the descent of *Juniperus wallichiana* into lower parts of these valleys, for in drier places in West Nepal this species can be found as low as 9,000 ft.

The increasingly dry conditions at the heads of these valleys are marked also by the occurrence of *Ephedra gerardiana* on stony riverside terraces and low mats of *Myricaria* on riverside gravel. The valleys have a pronounced glacial U-shape in their upper parts, and their broad flat floors are heavily grazed by yaks and sheep. Shrubs which are common here and which seem well adapted to withstand the grazing are *Hippophae thibetana*, *Juniperus wallichiana*, *Spiraea arcuata*, and *Berberis* species. Bushes of *Salix* grow along the streams. The low-growing species of *Myricaria* which occurs in these places is named in the *Flora of British India* as *Myricaria germanica* var. *prostrata*. It seems, however, that it should more correctly be named *Myricaria rosea*.

The glaciers descend to about 14,500 ft, and alpine shrubs grow on the terminal moraines and on the slopes above up to 15,500 ft, which is some 1,000 ft higher than alpine shrubs are normally found on the southern side of the main range. *Rhododendron nivale* is found only at the highest altitudes.

In the head of the Rolwaling valley between 13,500–15,500 ft I noted the following species:

> Juniperus wallichiana, Hippophae thibetana, Rhododendron anthopogon, R. lepidotum, R. nivale, Lonicera obovata, Lonicera myrtillus, Ephedra gerardiana, Potentilla fruticosa, Spiraea arcuata, and species of Salix, Berberis, and Cotoneaster.

Inner valleys west of Langtang

In contrast with the eastern parts of Nepal where only the heads of the inner valleys escape the monsoon rain, in the valleys of the western parts dry conditions often begin at much lower altitudes. In consequence several dry forest types can be found here. (See the sections dealing with *Pinus excelsa*, *Betula utilis*, *Cupressus torulosa*, *Juniperus wallichiana* forests.)

Within the country covered by these forest types there are slopes which at one time may have been forested but which now, probably due to a combination of fires, felling, and heavy grazing, are covered only with dry scrub. Since this scrub occurs at altitudes which lie within the forest zone it can hardly be classified as alpine, though in fact a number of the component species do also occur in the alpine zone. In the hard dry conditions prevailing here the distinction between alpine and non-alpine species is much less clearcut than in wetter parts of Nepal.

I have seen this dry scrub at altitudes within the zone of dry forests in many places on the fringes of the dry Dolpo–Mustang area. The shrubs do not much exceed 4 ft in height, and in the more exposed places they tend to form tight low-growing mats. I have noted the following species:

Juniperus wallichiana, Juniperus communis, Juniperus squamata, Lonicera hypoleuca, Lonicera myrtillus, Potentilla fruticosa, Spiraea arcuata, Ceratostigma ulicinum, Hippophae thibetana, Ephedra gerardiana, Rosa sericea, Sophora moorcroftiana, Caragana gerardiana, Plectranthus pharicus, Ribes alpestre, Rhododendron anthopogon, Clematis phlebantha, and species of Cotoneaster and Berberis. Artemisia species are abundant.

In the very dry windy valley of the Kali Gandaki around Jomsom the shrub flora at about 10,000 ft is much more limited; *Sophora moorcroftiana, Lonicera hypoleuca, Ephedra gerardiana,* and two species of *Caragana.*

Alpine steppes

To the north of the Dhaulagiri–Annapurna massif there lies some extensive steppe country with a climate and flora more typical of the Tibetan plateau than of the Nepal Himalaya. This country falls into three parts, Dolpo, Mustang, and Manang. Manang I have not visited, and Mustang because of the great rift valley of the Kali exhibits some rather special climatic and vegetational characteristics; I here confine my comments on the alpine steppe vegetation to that found in Dolpo.

That part of Dolpo which lies within the drainage of the Karnali headwaters has an extremely limited shrub flora. Between 13,500–16,500 ft two species dominate the open rolling hillsides to the almost complete exclusion of any others; *Caragana brevifolia,* a spiny plant up to 4 ft high, its yellow flowers having a bright chestnut wing and calyx; and the spiny pink-flowered *Lonicera spinosa.* For mile after mile at these altitudes these are the only two shrubs one sees in any quantity, except along the streams where there are thickets of *Salix* and *Myricaria.* In such few parts of this country as drop below 13,500 ft their dominance is broken, and they are largely replaced by *Caragana gerardiana.* Above 16,500 ft they give way to a shrubless cushion and scree flora. The *Myricaria* of the alpine steppes is a much taller plant than the almost prostrate *Myricaria* which occurs in the inner valleys of Central and East Nepal. It attains a height of 4–5 ft. It seems that it is a separate species, *Myricaria wardii.*

That part of Dolpo which lies within the drainage of the Bheri headwaters is rather wetter, for during the summer months quite an amount of rain filters up the Tarap khola and the Barbung khola. In consequence the shrub flora is rather richer. Species which I have noted here are as follows:

> Rosa sericea, Potentilla fruticosa, Rhododendron nivale, Rhododendron anthopogon, Lonicera obovata, Lonicera myrtillus, Spiraea arcuata, Juniperus wallichiana, and species of Berberis.

JUNIPERUS WALLICHIANA FOREST

Champion: No separate mention.

Before describing *Juniperus wallichiana* forest it will be as well to review the distribution of the juniper species which occur in Nepal.

Juniperus communis L.

This species is widespread in Europe and North Temperate Asia. It occurs in the NW. Himalaya but appears to be unrecorded from the Eastern Himalaya.

In Nepal it is confined to dry areas in the western half of the country, and even there it is not common. I have seen small shrubs of this species in open pine-birch forest at 10,500 ft in the Mugu Karnali valley; it has also been recorded from Jomsom; and the head of the Marsyandi valley.[23]

Juniperus recurva Buch.–Ham. ex D. Don.

This species ranges from Afghanistan through the Himalaya to Upper Burma and China. In Nepal it occurs in two different forms.

In tree form it commonly occurs between 10–12,500 ft as a component of the upper forest in all the wetter parts of the Midland areas. Sometimes it attains a height of 70 ft, though usually it is smaller. Forest clearings which face south or east are often colonised by this species, and in some places here one sees small patches of almost pure juniper forest which probably are of secondary origin.

In dwarf form it is very common in the alpine zone of the Midland areas. It ascends to 14,500 ft and often forms dense low mats. (See the section dealing with moist alpine scrub, p. 128.)

Juniperus squamata D. Don.

At one time this species was considered to be a variety of *J. recurva*, but I understand that it is now regarded as a distinct species. It ranges from Afghanistan through the Himalaya to Upper Burma and China. It is a dwarf alpine shrub, ascending to 14,500 ft.

In Nepal it has been recorded from the country west of the Karnali; from north of Dhorpatan; and from the inner valleys of West and Central Nepal. I have not seen this species further east than the Rolwaling valley, but it probably can be found in Khumbu or other of the eastern inner valleys. It appears to be absent from the wet southern slopes of the main ranges where *J. recurva* flourishes. In the dry country it is not nearly so common as *J. wallichiana*.

Juniperus wallichiana Hook. f. ex Brandis

This species has at one time been referred to as *J. pseudosabina* Hook., and in fact should correctly be called *J. indica* Bertol. I think, however, that *J. wallichiana* is the name most widely used. I discuss the distribution of the species below.

The Midlands

In the Midlands conditions generally are too wet for *J. wallichiana* to grow, but in the vicinity of Dhorpatan in the West Midlands local conditions are exceptional and it grows quite extensively, both as a tree and as a small shrub. The trees are 40–60 ft tall and in places here they form pure forest. The junipers are covered with lichen and yellow moss, and in the summer the damp grassy glades in the forest are bright with *Primula involucrata* and species of *Pedicularis*, *Anemone*, and *Polygonum*. This forest extends between 9,500–10,500 ft for some miles along the valley floor above Dhorpatan, and here I noted the following species:

1 Juniperus wallichiana. A few Abies spectabilis, Betula utilis, Quercus seme-carpifolia.

2 Rhododendron arboreum, R. campanulatum, Prunus cornuta, Prinsepia utilis, Lonicera lanceolata, Lonicera myrtillus, Rosa sericea, Rosa macrophylla, Syringa emodi, Prunus rufa, Viburnum cotinifolium, Cotoneaster acuminata, Jasminum humile, and species of Berberis and Arundinaria.
   Hippophae salicifolia and species of Salix and Myricaria along the streams.

3 Cotoneaster microphylla, Rhododendron lepidotum.

4 Clematis montana.

The climax forest on undisturbed ground here consists of *Abies spectabilis*. Perhaps the rather even-aged juniper trees of the forest described above are colonists of ground at one time cleared for grazing.

The Humla–Jumla area

J. wallichiana is quite common here as a shrub or small tree at altitudes between 9–11,000 ft in clearings in the *Pinus–Picea* forest. In these places it is obviously secondary in origin. I have not seen it forming forest here.

It is also common as a dwarf shrub in the alpine zone, largely replacing *J. recurva* at these altitudes.

Western inner valleys, and the edge of the arid zone

In the valleys which lead up to the dry steppe country lying to the north of the Dhaulagiri–Annapurna massif *Juniperus wallichiana* in some places occupies a dominant position in the vegetation. In the dry conditions which prevail there the principal tree species are limited to *Pinus excelsa, Betula utilis, Cupressus torulosa, J. wallichiana*, and sometimes a little *Abies spectabilis*. In some places an open type of steppe-forest composed of juniper develops, though often there is no true forest and the juniper trees are scattered singly across the dry rocky slopes. Their foliage looks very black against a background of bare hillside, and earns for the species the name of 'the black juniper'. Some of the trees may be 30 ft high, but the majority are small shrubs, and the distinction between this open type of forest and dry alpine scrub is not clearcut.

I have seen this type of forest in the Mugu Karnali, the Suli Gad, Tarap khola, Barbung khola, and upper Kali Gandaki. At 13,500 ft in a side valley of the upper Kali Gandaki which leads to the village of Sangdah there is some steppe-forest consisting of 20-ft trees of juniper, and I noted here the following species:

1 Juniperus wallichiana, Betula utilis.
2 Juniperus squamata, Spiraea arcuata, Caragana gerardiana, Potentilla fruticosa, Rosa sericea, Plectranthus pharicus, Lonicera spinosa, and species of Berberis, Salix and Cotoneaster.

No account of these steppe forests would be complete without mention of the aromatic smell of the vegetation. Even during the monsoon months the skies in these places are largely unclouded, and the hot midsummer sunshine on pine, cypress, juniper, and thyme produces a sharp dry smell very different from the wet heavy smell of the monsoon forests.

Eastern inner valleys

Here *Juniperus wallichiana* almost always occurs not as a tree but as a dwarf shrub (see the section dealing with dry alpine scrub, p. 129). In a few places, however, small trees of this species can be found occurring around the treeline, and I am told that in Simbua khola on the Sikkim border at an altitude of 12,500 –13,000 ft it forms almost pure forest over some acres with trees of 30–40 ft. In my experience this is unusual for East Nepal.

The groves of old and carefully preserved trees of juniper which surround some of the villages in Khumbu appear to consist exclusively of *Juniperus recurva*, although *Juniperus wallichiana* also occurs on nearby slopes as a small shrub.

PART III

Notes on Distribution

NOTES ON DISTRIBUTION

This study of distribution is limited to the species of trees, shrubs, and climbers mentioned in my descriptions of the forests of Nepal.

A wider study which included the herb flora would reveal far more endemic species, and many ruderals with a world-wide distribution. In so far as the vegetation of Nepal at altitudes below the treeline must at one time have consisted predominantly of forest a study limited to forest lists probably gives a fair picture of the distributional elements of which the original flora was composed.

The pattern of distribution shown by tropical and subtropical species differs so much from that shown by the temperate and alpine ones that it is necessary to consider the two groups separately, despite the fact that it is not always possible to draw a satisfactory dividing line between them.

A. TEMPERATE AND ALPINE FLORA

Before considering in detail the somewhat complicated division into various elements of the temperate and alpine flora of Nepal it may be helpful to summarise the position in outline.

1 The East Himalayan element is dominant in the flora as a whole. This element becomes much reduced as one travels westward through Nepal, but it has no terminal point there and carries on in diminished strength into the Western Himalaya.
2 Locally endemic species of trees and shrubs are not an important element in the flora.
3 The Tibetan element is dominant in the flora in some dry areas along the Tibetan border.
4 Some of the temperate species which occur in Nepal also occur in areas of high rainfall in the hills of South India, although they are absent from intervening parts of North and Central India.
5 The West Himalayan element is quite strongly represented in the flora of the western half of Nepal, particularly in the Jumla area and in some of the drier inner valleys.

   Many of the species of which this element is composed terminate their eastward range in the western half of Nepal, but some, although apparently absent from East Nepal and Sikkim, reappear further east.
6 A widespread north temperate element is present in Nepal. Some of the north temperate species extend their range throughout the whole length of the Himalaya, but a number of them terminate their range in the western half of Nepal in the same areas, and presumably for the same reasons, as many of the West Himalayan species.

   Some of these north temperate species, though apparently absent from East Nepal and Sikkim, reappear further east.

At the date of publication of the *Flora of British India* the Western Himalaya had been fairly fully explored, and so also had Sikkim. The collections of W. Griffith were available from Bhutan, but much of that country remained to be explored and Nepal too was only very imperfectly known from F.

Buchanan Hamilton's and N. Wallich's collections. The Assam Himalaya and SE. Tibet were entirely untouched. At that time therefore it was necessary to make the assumption that if a species had been recorded from the Western Himalaya but not from Sikkim it was West Himalayan in distribution and conversely that if it were known from Sikkim but not from the Western Himalaya it was East Himalayan.

Since then the flora of Nepal has been very fully collected, and there are also available in the herbarium of the British Museum (Natural History) the extensive collections made by F. Ludlow, G. Sherriff, F. K. Ward, and others from Bhutan, the Assam Himalaya and SE. Tibet, as well as those of G. Forrest, E. H. Wilson and others from Western China, and of Ward from North Burma. Although much remains to be done by further collecting in the field and more particularly by taxonomic botanists in the herbarium to correlate the Chinese and Himalayan floras we now have a much wider knowledge of the eastward range of many of the species included in the *Flora of British India*.

In the light of this knowledge two points become apparent. A number of species which previously were regarded as western in distribution are now known to extend their range not only into Nepal but in some cases, even though they may be unrecorded from Sikkim, into the extreme Eastern Himalaya. On the other hand a large number of species with an East Himalayan distribution, which had not previously been recorded west of Sikkim, are now known to occur in Nepal.

East Himalayan elements

1 *Species which as well as occurring in Nepal are widely recorded from other parts of the Eastern Himalaya; i.e. Sikkim, Bhutan, the Assam Himalaya, and SE. Tibet.*

None of these species have as yet been recorded from the Himalaya west of Nepal.

Acer campbellii
Acer caudatum
Acer hookeri
Acer pectinatum
Acer sikkimensis
Actinodaphne reticulata
Agapetes serpens
Aristolochia griffithii
Berberis insignis

Brassaiopsis glomerulata
Brassaiopsis hainla
Brassaiopsis mitis
Camellia kissi
Castanopsis hystrix
Corylus ferox
Daphne bholua
Diplarche multiflora
Edgeworthia gardneri

Ehretia macrophylla
Enkianthus deflexus
Euonymus porphyreus
Gamblea ciliata
Gaultheria fragrantissima
Gaultheria griffithiana
Gaultheria hookeri
Gaultheria pyrolifolia
Helwingia himalaica

Hoya fusca
Ilex fragilis
Ilex hookeri
Ilex intricata
Ilex sikkimensis
Larix griffithiana
Lindera heterophylla
Lindera neesiana
Litsea kingii
Litsea oblonga
Litsea sericea
Lithocarpus pachyphylla
Lithocarpus spicata
Lonicera acuminata
Lonicera cyanocarpa
Lonicera glabrata
Lonicera lanceolata
Maddenia himalaica
Magnolia campbellii
Magnolia globosa

Michelia doltsopa
Michelia velutina
Osmanthus suavis
Photinia integrifolia
Pieris formosa
Polygala arillata
Populus glauca
Prunus carmesina
Prunus rufa
Pyrularia edulis
Quercus lamellosa
Quercus lineata (sensu F.B.I.)
Rhododendron camelliae-
  florum
R. campylocarpum
R. ciliatum
R. cinnabarinum
R. dalhousiae
R. falconeri
R. fulgens

R. glaucophyllum
R. grande
R. hodgsonii
R. lindleyi
R. nivale
R. pendulum
R. pumilum
R. setosum
R. thomsonii
R. trichocladum
R. triflorum
R. vaccinioides
R. virgatum
R. wallichii
Ribes griffithii
Salix calyculata
Salix sikkimensis
Skimmia arborescens
Sorbus thomsonii
Stachyurus himalaicus

2 Species which as well as being widely recorded in the Eastern Himalaya and westwards through Nepal also occur in Kumaon–Garhwal and in some cases as far west as the Sutlej.

It seems that none of these species have been recorded from the Punjab Himalaya or from Kashmir. In view of their wide eastern distribution and the fact that they do not extend westward the full length of the Himalaya these species may reasonably be called East Himalayan.

Acanthopanax cissifolius
Acer sterculiaceum
Betula alnoides
Brassaiopsis aculeata
Castanopsis tribuloides
Cinnamomum tamala
Colquhounia coccinea
Coriaria nepalensis
Daphniphyllum himalayense
Dodecadenia grandiflora
Euonymus tingens
Eurya acuminata

Hoya lanceolata
Hydrangea anomala
Hydrangea heteromalla
Hymenopogon parasiticus
Jasminum dispermum
Juniperus wallichiana
Leucosceptrum canum
Leycesteria formosa
Lindera bifaria
Lindera pulcherrima
Litsea elongata
Lyonia villosa

Machilus odoratissima
Maytenus rufa
Neolitsea lanuginosa
Pentapanax leschenaultii
Piptanthus nepalensis
Prunus cerasoides
Prunus nepalensis
Prunus undulata
Quercus lanuginosa
Rhododendron barbatum
Ribes acuminatum
Rosa sericea

Sabia campanulata
Schefflera impressa
Schizandra grandiflora
Scurrula elata

Spiraea arcuata
Symplocos theaefolia
Toricellia tilliifolia
Tsuga dumosa

Viburnum cordifolium
Viburnum coriaceum
Viburnum erubescens

3 Species which as well as being widely recorded from the Eastern Himalaya and throughout Nepal range west of the Sutlej and into the Punjab Himalaya.

This range covers almost the whole length of the Himalaya, and in fact some of the species in this group continue on into Kashmir and across the Indus. I therefore refer to the group as pan-Himalayan.

Many of these species also range eastwards into Western China and for that reason should perhaps more properly be called 'Sino-Himalayan', but the problem of deciding whether species both of this group and of the two preceding East Himalayan groups should be considered as endemic to the Himalaya alone or as members of a wider Sino-Himalayan flora is often one which has not as yet been solved by the taxonomists.

I therefore do not feel justified in subdividing either the East Himalayan or the pan-Himalayan groups into Sino-Himalayan and Himalayan endemic sections. In view, however, of their close connections with species of the Chinese flora it can reasonably be assumed that most of these pan-Himalayan species have spread in from the east.

Abies spectabilis
Alnus nepalensis
Betula utilis
Carpinus viminea
Clematis connata
Cornus capitata
Cornus macrophylla
Cornus oblonga
Cotoneaster microphylla
Cynanchum auriculatum
Daphne papyracea
Deutzia hookeriana
Elsholtzia fruticosa
Gaultheria nummularioides
Gaultheria trichophylla
Hedera nepalensis

Holboellia latifolia
Ilex dipyrena
Jasminum humile
Juniperus recurva
Juniperus squamata
Lonicera angustifolia
Lonicera hispida
Lonicera myrtillus
Lonicera obovata
Lonicera quinquelocularis
Lyonia ovalifolia
Mahonia napaulensis (sensu
    F.B.I.)
Meliosma dilleniifolia
Microglossa albescens
Myrsine semiserrata

Neolitsea umbrosa
Pinus excelsa
Philadelphus tomentosus
Prinsepia utilis
Pyracantha crenulata
Pyrus pashia
Quercus glauca
Quercus semecarpifolia
Rhamnus virgatus
Rhododendron anthopogon
    (including R. hypenan-
    thum)
R. arboreum
R. campanulatum
R. lepidotum
Rhus semialata

Rhus succedanea	Sorbus foliolosa	Symplocos crataegoides
Ribes glaciale	Sorbus microphylla	Viburnum stellulatum
Rosa brunonii	Spiraea canescens	
Sorbus cuspidata	Spiraea bella	

From the above three lists even before considering the other elements of which the temperate flora of Nepal is composed it becomes evident that the East Himalayan element is the dominant one. Of a total of 254 temperate species under consideration 133 are unquestionably East Himalayan in distribution, and many also of the 55 pan-Himalayan species can be presumed to be eastern in origin.

Of the 133 species listed as East Himalayan 86 terminate their western range in Nepal. When travelling in East Nepal one cannot fail to notice the richness of the East Himalayan element present in the flora of the Arun–Tamur, and conversely one is impressed by the rapid disappearance of many eastern species as one moves west across the Arun–Kosi watershed. A similar abrupt contrast occurs between the vegetation of the hills north of Pokhara, where many East Himalayan species terminate their recorded western range, and that of the drier country to the west of the Kali Gandaki. Despite the vivid impression one receives in the field at these places of a sudden disappearance of many East Himalayan species it seems wrong to attach more than local significance due to local rainfall conditions to these rather abrupt variations in the composition of the flora. When viewed in the light of our knowledge of the distribution of the East Himalayan flora as a whole they appear to be no more than fluctuations in the gradual diminution of the eastern element which begins to the east of Nepal and which continues far beyond her western boundaries.

Of the 133 species listed as East Himalayan 47 continue on into Kumaon–Garhwal. When one bears in mind that many of the 56 species listed as pan-Himalayan also are eastern in origin it can clearly be seen that it is wrong to think that there is any terminal point in Nepal for the East Himalayan flora as a whole.

Locally endemic species

The species listed below are endemic either to Nepal alone or to Nepal and a short range of adjacent countries. The brevity of the list indicates that endemic trees and shrubs are not an important element in the Nepal flora.

It must be remembered, however, that the species here under consideration are only a small part of the Nepal flora, and if herb species had been included there would certainly be a much higher proportion of endemics.

Clematis phlebantha

This species has as yet been recorded only from the Jumla and upper Bheri areas of West Nepal.

Rhododendron cowanianum

This species has been recorded as far east as Khumbu and as far west as the Dhaulagiri range in East and West Nepal respectively.

Rhododendron lowndesii

This species has been recorded from the Upper Marsiandi, Tukucha, the Upper Bheri, and adjacent areas of Central and West Nepal.

Michelia kisopa

This species has been recorded as far west as Kumaon. In Nepal it is quite common, but it is said to be rare in Sikkim, and I can find no record of it occurring in Bhutan or further east.

Tibetan elements

Of the shrubs mentioned as occurring along the dry northern borders of Nepal a number are more Tibetan than Himalayan in distribution. Of these some are recorded from West Tibet and some from East Tibet, but the flora of that country is hardly well enough known to justify subdividing the list. The species are as follows:

Buddleja tibetica var. grandi-flora	Ceratostigma ulicinum	Lonicera hypoleuca
	Clematis orientalis	Lonicera spinosa
Caragana brevifolia	Ephedra gerardiana	Plectranthus pharicus
Caragana gerardiana	Hippophae thibetana	Sophora moorcroftiana

In these dry places at high altitudes shrubs are not a very important part of the flora, and the above brief list gives only a poor indication of the very strong Tibetan element in the flora here. Herb species with a predominantly Tibetan distribution are very numerous.

Temperate species which occur both in Nepal and in Peninsular India

It is somewhat surprising that certain temperate species occur both in the Himalaya and in the hills of South India. Much low-lying tropical country in which these species cannot survive intervenes between the two areas, so that how they succeeded in the past in crossing the gap is something of a puzzle. Various theories which have been put forward in explanation are discussed by G. S. Puri in his *Indian Forest Ecology*,[1] but the problem is more an Indian than a Himalayan one, and I only wish here to point out that there is this link between the flora of Nepal and of Peninsular India, and that of the above-mentioned trees and shrubs the following species occur in both areas:

Eurya acuminata	Jasminum officinale	Viburnum coriaceum
Gaultheria fragrantissima	Pentapanax leschenaultii	Viburnum erubescens
Jasminum humile	Rhododendron arboreum	

Species the distribution of which is doubtful

When plotting the distribution of the species mentioned in my descriptions of the forests of Nepal I have been surprised to find how rarely is there any doubt into which distributional category any particular species falls. Almost always there are sufficient field records to establish the range of the species, and such doubts as arise mostly are the result of difficulties of taxonomic distinction.

I do not feel justified in placing the following species in any category:

Cotoneaster acuminata, Cotoneaster frigida

The relevant material in the herbarium of the British Museum is on loan and unavailable. I think it probable that both species are East Himalayan in distribution.

Quercus incana

In view of the doubts mentioned in the section dealing with *Quercus incana–Quercus lanuginosa* forest about the eastward range of this species it seems better to omit it.

Machilus duthiei

The various species of *Machilus* are not easy to distinguish, and not all the gatherings from Nepal have as yet been named. I think it probable that this species is a West Himalayan one which ends its eastward range in East Nepal.

Larix potaninii

In the section dealing with *Larix* forest I point out that the only records of this West Chinese species of larch occurring anywhere in the Himalaya appear to be from Central Nepal. Such a distribution is not impossible, but it is so much at variance with the patterns of distribution shown by all other tree species that one suspects that the information here is incomplete.

Ilex insignis

It seems that the only Himalayan records of this species are from Sikkim and adjacent parts of East Nepal. Gatherings of this genus in the British Museum from Bhutan and SE. Tibet have not as yet been named, but H. Hara states[2] that it occurs in Yunnan, so it would seem probable that it does in fact occur in the Eastern Himalaya to the east of Sikkim.

Rhamnus purpureus

The Flora of British India gives this species a West Himalayan distribution, and there are no gatherings from Nepal from further east than longitude 83° 30'. It appears to be unrecorded from Sikkim, and gatherings of this genus in the British Museum from Bhutan and SE. Tibet are still unnamed. There is, however, one gathering of K. Ward's from the Burma–Tibet border named as this species, so it may yet be shown to have an East Himalayan as well as a West Himalayan distribution.

Viburnum grandiflorum

This species is recorded from Kashmir throughout the Himalaya as far east as Bhutan. It seems to be unrecorded from the Assam Himalaya, SE. Tibet, or Western China. If these records represent the true distribution of this species it does not fit into any normal distribution pattern, because with the exception of Viburnum cotinifolium and Hippophae salicifolia all the species with a wide West Himalaya range which manage to get as far east as Bhutan are also found further east again in SE. Tibet or the Assam Himalaya.

Lonicera purpurascens

This species is recorded from Swat and Kashmir as far east as Central Nepal. There are no records from East Nepal, Bhutan, Assam, or SE. Tibet, so that its distribution is essentially West Himalayan. There is however one gathering from Sikkim named as this species.

Euonymus fimbriatus

This species is recorded from Kashmir to Central Nepal. There are no records from East Nepal, Bhutan, Assam Himalaya, or SE. Tibet, but the Flora of British India states that it occurs in Sikkim.

The distribution of the above three species serves to remind us not to be too dogmatic about the precise place at which the West Himalayan flora terminates its eastward range.

Euonymus echinatus

This species is recorded from Kashmir to Sikkim, and there is a doubtful record from Assam mentioned in Kanjilal and Das's *Flora of Assam*. There are no records from Bhutan or elsewhere in the Eastern Himalaya, though with this pattern of distribution one would suspect that it in fact occurs there.

Wikstroemia canescens

This species is recorded from Afghanistan to Sikkim. I can find no record from Bhutan or SE. Tibet, but it is said to occur in Assam and there are gatherings from China.

With this pattern of distribution one would expect the species to occur in the Himalaya to the east of Sikkim. On the other hand in Nepal it certainly is much more abundant in the drier western areas than it is in the east, so perhaps it is in fact absent from the easternmost part of the Himalaya.

West Himalayan elements

1 *Species endemic to the Western Himalaya (and in some cases also to neighbouring parts of Afghanistan and Baluchistan) which end their eastern recorded range in Nepal.*

It is well known that the Western Himalaya contains an endemic flora much of which does not penetrate into the wetter Eastern Himalaya. Much of this flora has in fact disappeared before the western border of Nepal is reached, but a number of species continue on into the western half of that country where in the Jumla area and in some of the drier western inner valleys they constitute quite a significant part of the flora. The following list is limited to trees and shrubs; if it included herbs it would be very much longer.

Abelia triflora	Clematis barbellata	Skimmia laureola
Abies pindrow	Cupressus torulosa★	Sophora mollis
Aesculus indica	Morus serrata	Sorbus lanata
Alnus nitida	Picea smithiana	Staphylea emodi
Caragana brevispina	Pistachia integerrima	Syringa emodi
Cedrus deodara	Rhus wallichii	Ulmus wallichiana

★ But see note on distribution on p. 120.

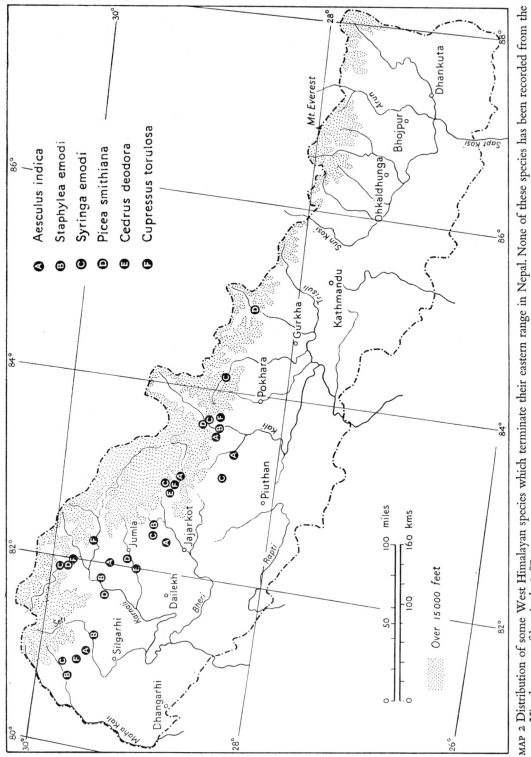

MAP 2 Distribution of some West Himalayan species which terminate their eastern range in Nepal. None of these species has been recorded from the Himalaya east of longitude 85°E.

These species appear to be able to prolong their range eastwards in Nepal by taking refuge in the drier inner valleys (see Map 2).

The termination of the range of these species, and of a number of other herb species not mentioned here, in the western half of Nepal gives support to the view that this part of the country represents the approximate terminal point in the Himalaya of the West Himalayan flora. We shall see, however, when we come to consider the next group of plants that this view, although containing a good deal of truth, cannot be accepted without qualifications.

2 *Species widespread in the Western Himalaya which are unrecorded from East Nepal and Sikkim, but reappear further east.*

The wetter conditions of East Nepal undoubtedly constitute a considerable barrier to the eastward advance of certain West Himalayan species. It seems, however, that some of these species, though absent from East Nepal and Sikkim, manage to cross this barrier and reappear in Bhutan. A number of species which reach Bhutan then continue their range eastwards for a very long way.

Meliosma pungens

Recorded from the Western Himalaya to Central Nepal. Not recorded from East Nepal or Sikkim. Recorded from Bhutan, SE. Tibet, China.

Plectranthus rugosus

Recorded from Baluchistan to Central Nepal. Not recorded from East Nepal and Sikkim. Recorded from Bhutan, SE. Tibet, W. China.

Populus ciliata

Recorded from the Western Himalaya to Central Nepal. Not recorded from East Nepal or Sikkim. Recorded from Bhutan, SE. Tibet.

Ribes emodense

Recorded from Kashmir to Central Nepal. Not recorded from East Nepal or Sikkim. Recorded from Bhutan, SE. Tibet, W. China.

Hippophae salicifolia

Recorded from Garhwal to Central Nepal. Not recorded from East Nepal or Sikkim. Recorded from Bhutan, but not SE. Tibet or China.

Viburnum cotinifolium

Recorded from Baluchistan to Central Nepal. Not recorded from East Nepal or Sikkim. Recorded from Bhutan, but not SE. Tibet or China.

All the above species have in common the fact that they reappear to the east having apparently been absent from East Nepal and Sikkim. The species listed below make a very much bigger jump; some disappear in the western half of Nepal and reappear in SE. Tibet, others are unrecorded nearer than West China.

Clematis grata

Recorded from Afghanistan to West Nepal. Unrecorded from Central Nepal eastwards in the Himalaya. *Clematis grata* var. recorded from China.

Lonicera webbiana

Recorded from the Western Himalaya to Central Nepal. Unrecorded from East Nepal, Sikkim, Bhutan. Recorded from SE. Tibet.

Olea cuspidata

Recorded from Baluchistan to West Nepal. Unrecorded from the Himalaya from East Nepal eastwards. *Olea cuspidata* var. recorded from W. China.

Quercus dilatata

Recorded from Aghanistan to Central Nepal. Unrecorded in the Himalaya from East Nepal eastwards. *Quercus dilatata* var. recorded from W. China.

Rhus punjabensis

Recorded from Kashmir to Central Nepal. Unrecorded from the Himalaya from East Nepal eastwards. *Rhus punjabensis* var. *sinica* recorded from W. China.

Sorbaria tomentosa

Recorded from the Western Himalaya to Central Nepal. Unrecorded from the Himalaya from East Nepal eastwards. Recorded from Central and North China.

The distribution of some of the above species is plotted on Maps 3 and 4. We can conveniently postpone an attempt to explain the rather strange break in the distribution of these species until we have considered the distribution of some north temperate species.

North temperate elements

1 Species widespread in the north temperate zone and throughout the Himalaya.

When studying the distribution of north temperate species it becomes very evident that the concept of a species is not fixed but fluid. Taxonomists often seem doubtful whether a plant should be regarded as part of a widespread north temperate species or should be split off as a separate Himalayan endemic one. Many plants which the *Flora of British India* refers to as north temperate species have subsequently been given separate specific status. *Hippophae thibetana* for example was previously lumped with *Hippophae rhamnoides; Prunus cornuta* with *Prunus padus; Potentilla arbuscula* with *Potentilla fruticosa.* There are many other cases, and as the study of the Himalayan flora proceeds there will doubtless be many more in the future.

This uncertainty as to whether the Himalayan form of plant is conspecific with the north temperate form is very often coupled with doubt as to whether the Chinese form is different again, and if so at which point in the Himalaya or in its extension eastward the Chinese form takes over from the Himalayan.

If one were to accept the old *Flora of British India* classifications there would be a number of widespread species common both to the north temperate zone and to the Himalaya. On modern classification of the limited number of species under consideration in these notes the only ones which seem to qualify under this heading are as follows:

Juglans regia, which is widespread in north temperate latitudes, is a common village tree in the Western Himalaya, and also occurs more rarely in the Eastern Himalaya. Its presence often is due to cultivation.

Jasminum officinale, which occurs both in China and in the Caucasus, has also been recorded throughout the Himalaya. In Nepal its presence also often seems to be due to cultivation.

2 Widespread north temperate species which terminate their eastward Himalayan range in Nepal.

Over a hundred years ago Sir Joseph Hooker had already divined that a number of north temperate species terminate their eastward Himalayan range somewhere in the western half of Nepal. He (or his co-author Thomson) wrote[3] 'One very remarkable result has already struck us with regard to the Himalayan

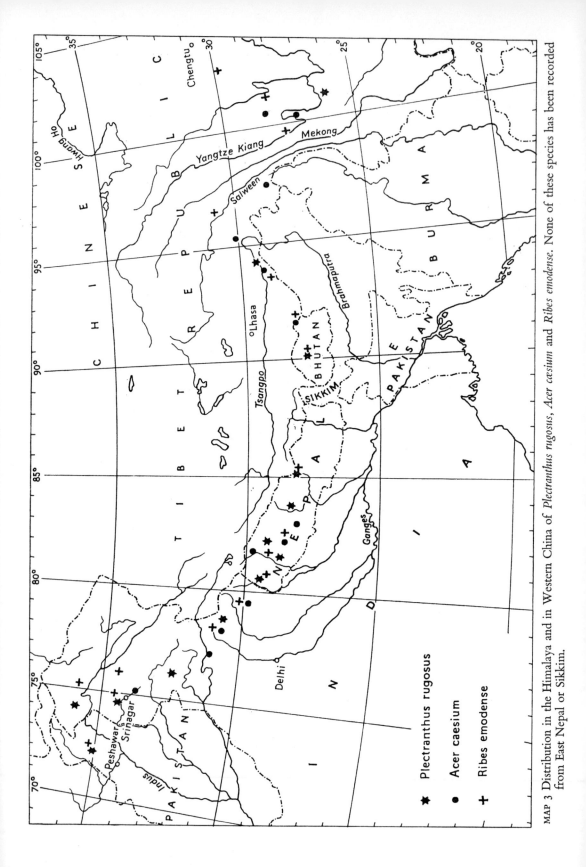

MAP 3 Distribution in the Himalaya and in Western China of *Plectranthus rugosus*, *Acer cæsium* and *Ribes emodense*. None of these species has been recorded from East Nepal or Sikkim.

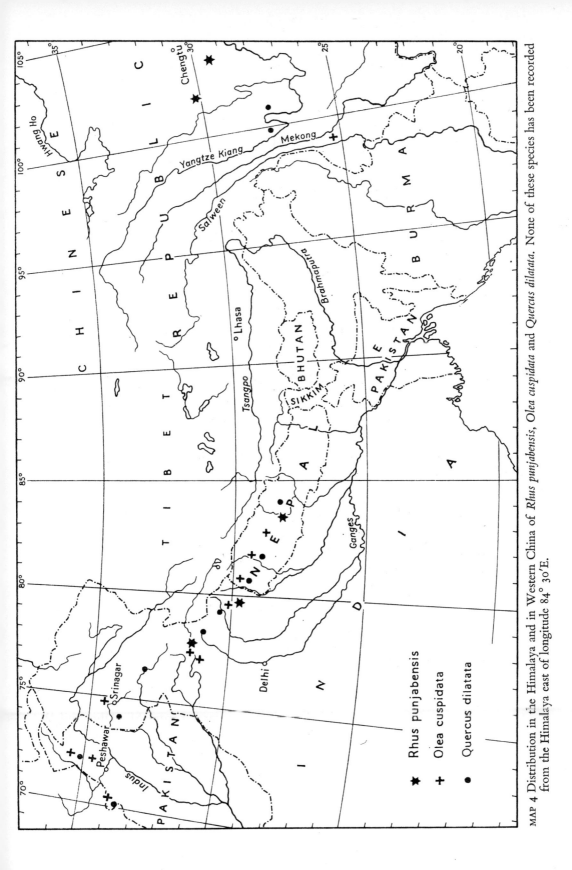

MAP 4 Distribution in the Himalaya and in Western China of *Rhus punjabensis*, *Olea cuspidata* and *Quercus dilatata*. None of these species has been recorded from the Himalaya east of longitude 84° 30'E.

distribution of European plants, namely their rapid disappearance to the east of Kumaon'.

Hooker made this statement at a time when knowledge of the Eastern Himalaya was virtually limited to Sikkim, and so he could not know that a number of the north temperate species which disappear in the western half of Nepal reappear again further east to form a pattern of distribution which closely resembles that which we have already seen developing for some of the so-called West Himalayan species.

Let us first consider the species which do in fact finally disappear in Nepal, and do not reappear further east. Of the limited number of species under consideration, the only one which fits into this category is *Juniperus communis*. This widespread north temperate species occurs in the Himalaya as far east as Central Nepal. It is known from Siberia and Mongolia, but seems to be absent from the Eastern Himalaya and Western China. The north temperate element, however, is far more strongly represented in the herb flora than amongst trees and shrubs, and if we were considering herbs we would see that there is ample evidence to justify Hooker's statement. For example, *Lathyrus luteus*, *Papaver rhoeas*, *Lamium album*, *Leonurus cardiaca*, *Gagea elegans*, and *Geum urbanum* all terminate their eastern recorded range in the western half of Nepal.

3 Widespread north temperate species which disappear in the western half of Nepal but reappear again further east.

Ribes alpestre, which occurs widely in the north temperate zone, is recorded from the Western Himalaya as far east as Central Nepal. I can find no records from the eastern half of Nepal or Sikkim, but it is known from Bhutan, SE. Tibet, and Western China.

The Himalayan distribution of the genus *Buxus* shows a similar pattern, though the situation is somewhat complicated by the fact that plants which the *Flora of British India* lumped with the north temperate *Buxus sempervirens* have now been split off. The West Himalayan form which is now known as *Buxus wallichiana* extends as far east as Central Nepal. There appear to be no gatherings of the genus from East Nepal or Sikkim, but specimens named as *Buxus sempervirens* have been collected from Bhutan, SE. Tibet, and West China. A separate species, *Buxus microphylla*, which is recorded from SE. Tibet and W. China, has also been recorded from Kumaon and West Nepal. Without concerning ourselves with the validity of the specific distinctions we can see

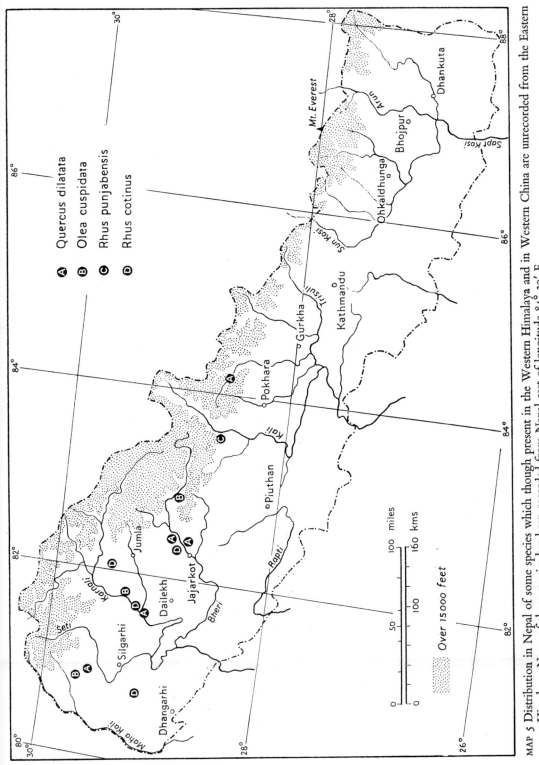

MAP 5 Distribution in Nepal of some species which though present in the Western Himalaya and in Western China are unrecorded from the Eastern Himalaya. None of these species has been recorded from Nepal east of longitude 84° 30′ E.

that there is a gap in recorded distribution of the genus in East Nepal and Sikkim.

The discontinuity in the distribution of *Rhus cotinus* is much bigger, equalling in distance that which we have already observed in the West Himalayan species *Quercus dilatata*, *Olea cuspidata*, and *Rhus punjabensis*. *Rhus cotinus*, which occurs in South Europe and West Asia, extends through the Western Himalaya as far east as Central Nepal (see Map 5). It appears to be unrecorded from anywhere east of this in the Himalaya, but *Rhus cotinus* var. *pubescens* occurs in West China.

Corylus colurna of Europe and West Asia occurs in the Western Himalaya as far east as West Nepal. In the Himalaya east of this it is replaced by *Corylus ferox*. *Corylus colurna* seems to reappear again in China, but as it is often cultivated and there are doubts as to specific distinctions its distribution had perhaps better be ignored.

Myrsine africana, although not a north temperate species, shows a similar gap in distribution. It ranges from East Africa through West Asia, Baluchistan, and the Western Himalaya as far east as Central Nepal. It is unrecorded from the Eastern Himalaya but reappears again in Central and North China.

We can see that the distribution of north temperate species in the Himalaya follows very closely the pattern already established for West Himalayan ones. There seems to be some obstacle in the eastern half of Nepal and in Sikkim to the further extension eastwards of all those species. Some species fail to cross the obstacle; others jump it to reappear in Bhutan; and some make a bigger jump to SE. Tibet or West China.

The East Nepal–Sikkim Gap

Possible reasons for the termination in the western half of Nepal of the range of a number of West Himalayan and north temperate species, and for the reappearance of certain of these species further east after a break in distribution.

Before seeking reasons for the above-mentioned patterns of distribution it would be as well to see whether the gap which has become apparent in East Nepal and Sikkim is in fact a gap in distribution or merely a gap in records.

The temperate flora of Sikkim and the neighbouring Chumbi valley has been very thoroughly collected, and the group of plants which appear to be absent here is too big for one to believe that any significant part of it can have been overlooked from lack of collecting.

East Nepal has not been subjected to quite such intensive collecting as Sikkim, but by now its flora is pretty well known. I myself have travelled in most parts of East Nepal, and I have seen that, apart from a few restricted areas in big river valleys which are dry enough to enable *Pinus roxburghii* to grow, there is no really dry country here until one has ascended some of the inner valleys such as Khumbu and Kambachen to altitudes above the treeline. In the section dealing with inner valleys I have pointed out that whereas in the western half of Nepal the rainfall in the inner valleys begins to fall off at altitudes sufficiently low to permit the occurrence there of various types of dry forest this is not so in the eastern half of the country. In the east there is no significant falling off of rainfall in the inner valleys until an altitude of about 13,500 ft is reached, so that dry types of forest are not found here. The missing species are all ones which are widespread in the drier Western Himalaya, and it seems probable that they are absent from East Nepal and Sikkim because of a lack of sufficiently dry habitats.

The Tibetan side of the border has yet to be fully explored, but in many places here country on the southern sides of the main ranges with a heavy monsoon rainfall abuts onto high dry Tibetan plateau country without the intervention of any medium dry country at altitudes suitable to provide habitats for the missing species. Some of these species may in due course be found in places such as Kyerong, the Rongsha Chu, and the Karta Chu, but I do not think further exploration in these quite limited areas will make any very great difference to the general picture of the gap in distribution as shown by present records. When one reflects upon the matter one can see that it is not the existence of this gap which is surprising. It would be surprising if the gap were shown not to exist, for the undisputed absence from the Eastern Himalaya of so many of the West Himalayan and north temperate species absolutely demands that there should be some such gap reflecting the existence of a climatic barrier to the further eastward advance of these species.

I have already indicated that I think the primary reason for the absence of these species from East Nepal and Sikkim is the absence there of sufficient medium dry habitats below the treeline. The Western Himalaya with its generally lower rainfall has plenty of these habitats. Increased rainfall and lack of suitable habitats seem sufficient in themselves to explain the termination in the western half of Nepal of much of this western flora, but before going on to consider why some of it reappears further east I must point out that certain other factors may also be operating to prevent its eastward extension.

The western half of the Himalaya runs in a north–west to south–east direction,

so that between Srinagar in Kashmir and Gangtok in Sikkim there is a difference of about seven degrees of latitude. From Sikkim there is a slight trend
northwards, so that the bend of the Tsangpo in SE. Tibet is some three degrees
north of Gangtok.

As a result of the difference in latitude length of daylight varies much more
between summer and winter in Kashmir than in Sikkim. Also the winters are
very much colder in the north. For example, heavy snowfalls are common-
place in Srinagar, but in Kathmandu, which lies at similar altitude but lower
latitude, they are unknown.

Winter conditions in the Western Himalaya therefore have a closer resemblance to a north temperate climate than do those in East Nepal and Sikkim.
This fact coupled with changes in length of daylight may be in part responsible
for the fading out of much of the north temperate flora from East Nepal and
Sikkim. In my opinion the dominant reason for their absence is that they are
driven out by a lack of sufficient dry habitats, but it is as well not to view the
problem too exclusively in terms of heavy summer rainfall.

Let us now pass on to consider why certain species which disappear in East
Nepal and Sikkim reappear again further east. Before considering Bhutan, let
us look at the country where SE. Tibet adjoins the Assam Himalaya. The Assam
Himalaya undoubtedly receive a very heavy rainfall, probably heavier than that
received by East Nepal and Sikkim.

In the latter areas, however, the Himalayan range is at its highest and acts as a
very effective rainscreen which abruptly separates the wet southern sides from
the dry northern ones, whereas in the Assam Himalaya the main mountain
chain with the exception of isolated peaks such as Namcha Barwa is very much
lower and less continuous, and therefore the change from wet to dry is much
less abrupt. F. Ludlow[4] puts the matter thus:

'From Kashmir to the Assam Himalaya, as far east as longitude 93, the main
Himalayan axis acts as an effective rainscreen. It separates, very abruptly, a wet
forested area in the south from a dry desert area to the north. Each of these two
areas has its characteristic avifauna, and the boundary is so sharply defined that,
in crossing the main range, one passes from one into the other generally in the
course of a day's march. It is only where large rivers cut through the main
range that we ever find forested country north of the main axis, and this is
neither extensive nor dense.

'But east of longitude 93 there comes a change. The main range here sinks down to comparatively low levels, so that for 120 miles there are no peaks of any magnitude, until suddenly Namcha Barwa (25,445 ft) rears its great height to the sky, girdled by the Tsangpo. The passes across the main range between the 93rd and 95th meridians probably do not average more than 13,500 ft, whilst some are as low as 12,000 ft with conifers growing on their summits. . . . The main range therefore, between the 93rd and 95th meridians, although undoubtedly the first line of defence against the onslaught of the monsoon, is no longer the Maginot line it is west of the 93rd meridian. But other ranges north of the main range act as secondary rainscreens and although the greater part of the Gyamda and Yigrong valleys in Kongbo and Pome are well wooded, they become progressively drier, and in their upper reaches lie within the dry zone. The rainscreen between the 93rd and 95th meridians is therefore not sharply defined, and is fanned out northwards.'

Here in SE. Tibet it seems that there is quite an extensive area of country covering the secondary ranges north of the main Himalayan divide where a diminished rainfall provides the medium dry conditions favoured by the species missing from East Nepal and Sikkim. Ludlow in the same work and with the help of information supplied by Kingdon Ward goes on to trace the rainscreen eastwards into Western China, and it is evident from the map provided that here too there are areas where the monsoon rains fall in gradually decreasing amounts over wide ranges of country.

Kingdon Ward himself was aware of the reappearance here in this drier zone of some of the West Himalayan flora, for in 1936 he wrote, 'It is interesting to observe that many species of plants found in the dry north-western Himalaya, but unknown in the eastern Himalaya, are found on the outer plateau north of the Assam Himalaya, and even north of the Tsangpo.' He does not enlarge on this point, nor specify any of the plants concerned.[5]

An explanation to the problem of the missing species therefore can be put in crude but I think more or less plausible terms thus:

1 The monsoon rainfall in the Western Himalaya is sufficiently low to permit the missing species to grow on the southern sides of the main ranges.

2 By the time West Nepal is reached the increased rainfall has driven these species into the drier inner valleys.

159

3 In East Nepal and Sikkim rainfall even in the inner valleys does not fall off greatly until altitudes above the treeline are reached. The northern sides of the main ranges here change abruptly to a high dry plateau climate unsuitable for these species, which are therefore squeezed out between very wet and very dry conditions.

4 Many species both of the West Himalayan and of the north temperate floras terminate their eastern range where the squeeze is at its most intense. Some reappear again east of the 93rd meridian where the squeeze has been relaxed.

This explanation does nothing to solve the further problem of how certain species have succeeded in crossing from one area to the other, but consideration of this point can conveniently be deferred until we have considered Bhutan.

Monsoon rainfall in Bhutan in general is probably as heavy as it is in Sikkim, and although the Bhutan section of the Himalaya does not contain any giants as big as Everest or Kanchenjunga, the main crestline is still very high and in consequence the transition from monsoon conditions on the south side to dry conditions on the north is abrupt. This combination of facts does not seem likely to produce the medium dry conditions favoured by the species with which we are concerned. Despite this, however, of the 17 species I have listed as jumping the East Nepal–Sikkim gap 8 are recorded as reappearing in Bhutan, and constitute strong evidence that some dry areas must occur within that country.

Additional evidence is provided by some other facts. *Pinus roxburghii*, which is a good indicator of drier conditions, is said to be abundant in some parts of Bhutan. This species is rare on the Teesta, and although present on the lower Arun and Tamur in limited amounts it is almost completely absent from their wet upper parts. *Pinus excelsa* and *Quercus semecarpifolia*, both of which are absent from Sikkim,[6] reappear again in quantity in Bhutan. And finally there is here in Bhutan by far the biggest area of settlement on the southern side of the Himalaya of people of Tibetan stock and culture who normally avoid areas of heavy rainfall. From all these facts one can deduce that there must be at least some parts of Bhutan which are not so wet.

Local topography can have a remarkable effect on reducing local rainfall even in country with a high general rainfall, and it seems that this is the reason why some parts of Bhutan are very much drier than others. Kingdon Ward makes the following comment: 'From the Teesta, which forms the valley of Sikkim, eastwards to the Subansiri in Assam, no large river drains the Eastern Himalaya. Consequently the monsoon has no direct access to the interior valleys. It is compelled to make its way by frontal assault over the outer range instead of through a wide breach in it, as in Sikkim and NE. Assam. It is this structure which gives to Bhutan, and to a short length of the Assam Himalaya, their peculiar character; peculiar that is for the Eastern Himalaya in general. For the valleys which lie immediately behind the outer range are completely dry.'[7]

If this statement is correct it explains the existence of dry areas in Bhutan. It does, however, seem to ignore the existence of the great Manas river which cuts through from the Tibetan side of the frontier to the plains. Without personal knowledge of Bhutan I cannot testify as to the accuracy of the rest of his remarks, but it is relevant to point out that although Ward knew the Eastern Himalaya well he passed through Bhutan only briefly on one occasion.

Ludlow, who has travelled extensively in Bhutan, puts the matter rather differently. He states, 'Paradoxical as it may seem, there are curious dry areas in Bhutan. They may be found in some of the major valleys of the interior between 3–6,000 ft. In these areas the hill slopes are almost destitute of forest, and exhibit a xerophytic type of vegetation. It is difficult to account for this dry river zone. Possibly it may be due to a current of hot air sucked up through the river gorge from the plains. . . . But the area of this dry zone is sharply circumscribed, and an ascent, sometimes of a few hundred feet, is sufficient to transport the traveller from comparatively arid country into dripping rain forest.'[8]

Similar dry river valleys exist in Nepal, particularly on the Karnali and Bheri, but here the general rainfall prevailing in the surrounding countryside is lower than in Bhutan, so that the contrast in vegetation over a small distance is not so dramatic. Ludlow states that these dry places occur between 3–6,000 ft, so that they would account for the presence of *Plectranthus rugosus*, and probably account too for the extensive forests of *Pinus roxburghii*, but they do nothing to explain the presence of species such as *Ribes alpestre* and *Ribes emodense*, both of which in Nepal are plants which occur in the drier types of upper forest. With the occurrence in Bhutan of these species, and also of species such as *Larix griffithiana* and *Picea spinulosa*, I suspect that despite an overall heavy rainfall some of the heads of the inner valleys there must show a falling off in rainfall at altitudes sufficiently low to permit the development of rather drier types of forest below the treeline. If this be so it must be caused by the local topography of the country.[9]

From all the foregoing facts it becomes apparent that although in the past the Himalaya has been an effective corridor for the westward migration of many Sino-Himalayan species it has been a very much less effective corridor for the eastward migration of West Himalayan and north temperate species. The reasons why so many of the latter species do not penetrate into the Eastern Himalaya have been discussed above; it is now time to consider how some of these species have jumped the gap.

The distance from Central Nepal to Bhutan is about 350 miles; to SE. Tibet about 750 miles; and to the relevant parts of Western China about 1,000 miles. The gap with Bhutan is not so great that it could not have been bridged by normal means of dispersal, but when one considers the gap between the Nepal trees of *Quercus dilatata, Olea cuspidata, Rhus punjabensis,* and *Rhus cotinus* and their nearest relations in Western China a thousand miles away to the east one begins to wonder if some other force has been at work. In the section of the Himalaya where the gap in distribution occurs two totally different types of climate to north and south of the mountains are in proximity to each other, so it seems possible that in the past quite small changes in climate or topography might have triggered off considerable changes in local rainfall. All that would be required to link up the eastern and western areas of distribution would be a narrow corridor of country at medium altitudes with medium rainfall.

The existence of such a corridor in the past is a mere speculation, but if in fact it did exist it would conveniently solve another problem. Writing of *Deutzia hookeriana* H. K. Airy Shaw states, 'The segregation of this species from *Deutzia corymbosa* affords yet another example of the distinctness of the East and West Himalayan floras. The occurrence of pairs of allied vicarious species in these two regions is probably more frequent than that of one species common to both.'[10] Anyone who knows the Himalayan flora will appreciate the truth of that remark. Examples are *Abies pindrow, Abies spectabilis; Picea smithiana, Picea spinulosa*. I think it is generally assumed that these pairs, in each of which one species is closely allied to the other, must in the past have evolved from a common stock, and that even though some of these paired species now overlap in range each separate half of the pair must have evolved on separate lines at a time when they were geographically separate.

Let us consider the genus *Picea*, which still remains split in two. *Picea smithiana*, the West Himalayan spruce, extends east as far as longitude 85. *Picea spinulosa*, the East Himalayan spruce, extends west as far as longitude 88. There is little doubt that much of the West Himalayan flora must have entered the area from the west; for example, with the present distribution of other members of the genus *Cedrus* it can hardly be supposed that *Cedrus deodara* entered from any other direction. If, however, we assume that *Picea smithiana* entered from the west and *Picea spinulosa* from the east we are confronted with the supposition that two closely allied species entered the Himalaya from different ends which are 2,000 miles apart. Since their close relationship presupposes evolution from a common stock we are also faced with the need to find a point of origin, presumably somewhere to the north, where this stock at one time grew. Although this solution is not impossible it requires the two species to have moved great distances by roundabout routes, and these movements can only have taken place under climatic conditions very different to those of the present day.

If, instead of supposing that these paired species entered the Himalaya from different ends, we imagine they were once part of a continuous flora stretching without a break the whole length of the Himalayan range, we can see that only quite small changes in climate would have been needed for them to have reached their present divided positions. We should need to imagine that at one time the central part of the Himalayan corridor was more suitable for the passage of species which prefer a medium rainfall than it is today. Then if subsequently the corridor was closed by some small shift in climatic conditions or by the further elevation of the mountain chain to form a more effective rainscreen the drier flora would have been split in two, and would have continued to evolve on separate lines. Finally at some later stage it seems possible that some of the eastern species, presumably because in the course of evolution they became more adaptable to wet conditions, managed to cross the gap along the wet southern side of the mountains and reinvade the Western Himalaya. This would account for the fact that certain linked species such as *Abies spectabilis* and *Abies pindrow* now overlap each other in range. The western species of these pairs have been less successful in their advance eastwards.

Although this solution is purely speculative it has the attraction of also explaining why certain West Himalayan species reappear far away in Western China. Whether the Chinese specimens of *Quercus dilatata*, *Rhus punjabensis*, *Rhus cotinus*, and *Olea cuspidata* are indeed conspecific with their western counterparts or whether at some time in the future taxonomists will give them separate specific status seems to be immaterial. They are sufficiently similar for one to be compelled to believe that they once formed part of one continuous flora.[11]

B. TROPICAL AND SUBTROPICAL FLORA

The elements which compose the tropical and subtropical flora of Nepal can be summarised briefly as follows:

1 The dominant element is a widespread North Indian one, mainly composed of deciduous trees which are well adapted to a climate with a markedly seasonal rainfall.

2 There is also an Assam–Burma–Malaya element present in the low-altitude flora. This element consists mainly of evergreen species, and is confined to special situations. It is well represented in the extreme east of the country, but becomes progressively more impoverished as one moves westwards.

   Some of these evergreen species, though absent from much of North and Central India, reappear in areas of high rainfall in the hills of South India.

3 Locally endemic species are few.

4 The Western Himalaya lie too far north to have developed a truly tropical or subtropical flora. Certain species which occur in hot low-altitude areas of the Western Himalaya and Baluchistan also occur in Nepal, but they are confined to unusually dry situations.

Eastern elements

We have seen that in the temperate flora of Nepal the predominant element is an eastern one which extends westwards in decreasing strength through Nepal and on into the Western Himalaya. A similar eastern element is detectable in the tropical flora, but it differs from the temperate one in two ways:

1 Its connections are much more with the flora of Assam–Burma–Malaya

than with the Sino-Himalaya flora, though in so far as subtropical species are not easily separable from lower temperate ones these floras to a certain extent overlap each other.

2 It is not predominant in the tropical flora of Nepal as a whole, but only in evergreen and semi-evergreen types of forest which occur in the eastern half of the country. It is poorly represented in the more widespread deciduous types of forest.

Many eastern species which occur in Assam have already disappeared from the foothills before the borders of Nepal are reached, and undoubtedly this type of flora is ill adapted for survival in a climate with a long dry season. Of the species which do reach Nepal many appear to terminate their westward range in the extreme east of the country, between the Kosi and the Sikkim border, and mostly they are confined to localities where conditions are unusually damp and shady. The general picture is of a flora which is only just able to survive in Nepal and which decreases in numbers very rapidly as one travels westwards.

A study of the composition of tropical evergreen forest illustrates the latter point. Of the 71 species which I mention as occurring in this forest type, 43 are unrecorded from Kumaon or the Western Himalaya. They are as follows:

Acrocarpus fraxinifolius	Dysoxylum procerum	Mucuna macrocarpa
Actinodaphne angustifolia	Ehretia wallichiana	Myristica kingii
Actinodaphne obovata	Garcinia xanthochymus	Myristica linifolia
Albizzia lucida	Gynocardia odorata	Ostodes paniculata
Antidesma acuminata	Lithocarpus spicata var. brevi-	Poikilospermum lucidum
Aphania rubra	petiolata	Polypodium coronans
Baccaurea sapida	Litsea oblonga	Pseuderanthemum indicum
Carallia lucida	Machilus villosa	Sarauja roxburghii
Castanopsis indica	Macropanax undulatum	Solanum crassipetalum
Cleidion javanicum	Maesa chisia	Sterculia coccinea
Clerodendrum nutans	Mangifera sylvatica	Trevesia palmata
Cryptocarya amygdalina	Meliosma simplicifolia	Vernonia talaumifolia
Cycas pectinata	Melodorum bicolor	Vernonia volkameriaefolia
Duabanga sonneratioides	Micromelum integerrimum	Villebrunea integrifolia
Dysoxylum binectariferum	Morinda angustifolia	

The composition of other types of evergreen and semi-evergreen forest which occur at low altitudes in the eastern half of Nepal confirm the view that the eastern element in the flora falls off rapidly as one moves westwards. Of the species mentioned in my descriptions of subtropical semi-evergreen forest, subtropical evergreen forest, and *Schima–Castanopsis* forest the following species appear to terminate their westward range in Nepal and to be absent from Kumaon and the Western Himalaya.

Acer thomsonii	Ecdysanthera micrantha	Ormosia glauca
Alstonia neriifolia	Eugenia ramosissima	Oxyspora paniculata
Ardisia macrocarpa	Eugenia tetragona	Phoebe attenuata
Capparis multiflora	Evodia fraxinifolia	Picrasma javanica
Clerodendrum bracteatum	Helicia erratica	Podocarpus neriifolius
Clerodendrum colebrooke-	Laportea sinuata	Randia fasciculata
anum	Luculia gratissima	Sapium baccatum
Croton caudatus	Macaranga denticulata	Schima wallichii
Cyathea spinulosa	Mallotus nepalensis	Talauma hodgsonii
Debregeasia wallichiana	Miliusa macrocarpa	Wightia speciosissima
Dichroa febrifuga	Mezoneurum cucculatum	Zizyphus incurva
Dobinea vulgaris	Myrsine capitellata	
Drimycarpus racemosus	Natsiatum herpeticum	

It is therefore evident that many eastern tropical and subtropical species fade away in Nepal and do not reach the Western Himalaya. I think, however, it would be equally wrong to regard Nepal as being the precise terminal point of this type of flora as it is to regard it as the terminal point of the Sino-Himalayan temperate flora. In both cases much of the eastern flora has faded out before the borders of Nepal are reached; and in both cases some eastern elements pass right through Nepal and on into the Western Himalaya. From the subtropical flora, for example, both *Castanopsis tribuloides* and *Pandanus furcatus* have been recorded from Kumaon; and *Phoebe lanceolata* is known as far west as the Beas.

NOTE: *Eastern elements which occur in South India*

It is rather curious that a number of species of the Assam–Burma–Malaya flora which occur in the Himalayan foothills are absent from the North Indian plains but reappear in areas of heavy rainfall in the coastal hills of Peninsular India. The problem of how these species reached south India is an interesting one, but is more relevant to a study of Indian than of Nepalese vegetation.

Widespread North Indian elements

The tropical evergreen forests of Nepal, even though very limited in extent, are so rich in eastern species that they are a very rewarding field for the collector and tend to claim an undue proportion of his time. Although less interesting to the collector, the various types of deciduous tropical forest are in fact far more widespread in Nepal, and their component species are far more typical of the general tropical vegetation of the country. These species are almost all widespread not only in Nepal but in North India as well. Many of them occur in South India, and most of them extend along the Himalayan foothills far to the west of Nepal.

The latter point is illustrated by the composition of sal forest. Of the 39 species mentioned in my descriptions of this type of forest, 33 occur also in Kumaon and the Western Himalaya. There is thus a big contrast between the widespread flora of this sal forest as compared to that of tropical evergreen forest, of which over half the component species terminate their westward range in Nepal. The sal forest species which do not occur west of Nepal are as follows:

Terminalia myriocarpa, Dillenia pentagyna, Amoora decandra, Croton oblongifolius, Clausena excavata, and *Aporosa dioica.*

Because it is so widespread, the flora of the tropical deciduous forests, as well as being not particularly interesting for the collector, is not of much concern to a student of distribution of the Nepalese flora. As regards this type of flora Nepal is merely an extension of North India. These widespread North Indian species are well adapted to a climate with a markedly seasonal rainfall, but in Assam where rainfall is heavier and the dry season shorter they are to a large extent replaced by an evergreen type of flora. In the foothills and subhimalayan tracts of Nepal this changeover from one type of flora to the other is only just beginning to take place. Even in the extreme east of the country evergreen forest is limited to special situations. The process of change from one flora to the other seems to quicken very much in the foothills of Sikkim and Bhutan.

Locally endemic species

We have seen that the great majority of species which compose the tropical and subtropical vegetation of Nepal have a wide distribution elsewhere in India, and in some cases in other countries beyond. Of the species mentioned in my descriptions of the vegetation the only ones with a more limited distribution are as follows:

Piptadenia oudhensis

This species is known only from Kumaon, the extreme west of Nepal, and the adjacent forests of Oudh.

Homalium nepalense

The *Flora of British India* gives the distribution of this species as Nepal alone, but it is now recorded also from Bengal, Orissa, and Chota Nagpur. As it is quite common in eastern parts of Nepal it seems curious that there should be no records from Sikkim or further east.

Leucomeris spectabilis

This species is recorded from Kumaon to East Nepal. There appear to be no records from Sikkim and Bhutan, but since there is one record from Kamroop in Assam it seems possible that this species does in fact occur in the Eastern Himalaya.

Eriobotrya elliptica

This species is recorded from Kumaon to East Nepal, but not from Sikkim, Bhutan or Assam where it seems to be replaced by *Eriobotrya petiolata*.

Eugenia frondosa

This species is recorded from Kumaon to East Nepal. I can find no records from Sikkim or Assam, but there is one doubtful identification from Bhutan. This genus is not an easy one, and specific identifications may not always be reliable.

It seems that Nepal and the Central Himalaya do not possess an endemic tropical forest flora. Of the above species the only one which is unquestionably a very local endemic is *Piptadenia oudhensis*.

Western elements

Although no part of Nepal lies within the tropics one is justified by convention and convenience in referring to its low altitude flora as tropical, for many of the species of which this flora is composed extend their range into the tropics, and the general character of much of the vegetation is similar to that found in tropical India. The western end of the Himalaya lies a long way north of Nepal, and a truly tropical vegetation is not found there. *Dalbergia sissoo*, which is closely associated with water and thus better able to survive in the dry north-west than most other tropical species, ranges the full length of the subhimalayan tracts and into Baluchistan, but in general the low altitude vegetation in these dry places is scrub, often of a thorny nature, which one can only rather doubt-fully describe as subtropical. This type of vegetation has to contend with intense heat and drought in summer and considerable cold in winter.

Very little of this western low altitude flora reaches Nepal. Presumably it does not thrive in summer monsoon rainfall conditions, and such little of it as occurs in Nepal is confined to special situations such as dry river valleys. Unlike the great majority of subtropical species which must have colonised Nepal from south and east the following plants are outliers from the west: *Olea cuspidata, Plectranthus rugosus, Pistachia integerrima, Zizyphus oxyphylla* and *Rhamnus triqueter*. All occur in Nepal, and have a range westwards through the Himalaya to Afghanistan and Baluchistan.

Capparis spinosa, which occurs in dry valleys in West Nepal, ranges westwards through Baluchistan and on to the Mediterranean.

Pinus roxburghii has a range from Afghanistan through to Bhutan, but since it is most extensively found from West Nepal westwards it can reasonably be regarded as a West Himalayan species. It does not fit easily into either a sub-tropical or a temperate category, because although it occurs principally between 3–6,000 ft it can on occasions be found as high as 9,000 ft.

Collecting Journey Routes

1954

In this year I collected as a member of the expedition jointly sponsored by the British Museum (Nat. Hist.) and the Royal Horticultural Society. The expedition was in the field between 11 April and 4 November. Collections were made under joint Stainton, Sykes and Williams numbers.

I myself travelled in April and May westwards from Pokhara to Baglung and Dhorpatan, and thereafter I spent the rest of the season in the area of Ghasa, Tukucha, Muktinath, and Mustang.

1956

In this year I collected in East Nepal. I started from Dharan on 16 April, and spent May and June collecting on the upper Arun. I visited Num, Topke Gola, Thudam, Chepua, and the Barun and Cchoyang valleys. In July and part of August I travelled from Num to Topke Gola, Thudam, Walungchung, Yangma, Kambachen, and Taplejung, returning to Num via Topke Gola. In late August I visited the Popti La on the Tibetan frontier, and in September I revisited the Barun valley before returning to Dharan.

I made 2,097 gatherings of plants.

1962

In this year I collected in Central Nepal between 10 April and 10 August. For part of this period I was accompanied by S. A. Bowes Lyon who collected under separate numbers.

I visited Langtang, Chilime, and Satsae khola to the south of Ganesh Himal, during April and May. At the beginning of June I returned to Kathmandu via Gosainkund and Malemchi. In June and July I travelled via Satsae khola to Prok on the upper Buri Gandaki, returning thence to Kathmandu via Chilime, Langtang, and the Ganja La.

I made 528 gatherings of plants.

1963

In this year I collected in West Nepal between 7 April and 30 July. Starting from Nepalganj I travelled by way of the Rapti valley, Dang, and Jajarkot to Kaigaon on the Jagdula khola. In May I made a circuit from Kaigaon to Jumla, Rara, and Maharigaon. In June I travelled eastward to Tibrikot, Dunaihi, and the lake at Ringmo. Thence in June and July I made a circuit through Dolpo via Sya Gompa, Phijor, Ccharka, and Tarap, returning to Dunaihi. From here I returned to Pokhara via Mukut and Tukucha.

I made 247 gatherings of plants.

1964

In this year I collected in Central and East Nepal between 5 April and 26 July. I was accompanied by D. McCosh for the whole of this period and by S. A. Bowes Lyon for part of it. We collected under separate numbers.

From Kathmandu we travelled to Junbesi. In April and early May I visited Chaunrikharka, Thyangboche, the Inukhu khola, and the Salpa Bhanjyang. In late May and June I collected at the head of the Solu, Likhu, and Khimti kholas and revisited Chaunrikharka. In July I visited Rolwaling, returning to Kathmandu via Jata Pokhari, Panch Pokhari. and Jiri.

I made 350 gatherings of plants.

1965

In this year I made two separate journeys.

Between 17 April and 1 July I collected in the extreme west of Nepal with Tirtha Bahadur Shrestha of the Department of Medicinal Plants. We started from Dhangarhi and travelled via Silgarhi Doti and Kaptad to Chainpur on the Seti river. From there we made a circuit in early May westwards to the Kali Gad, and in late May I made another circuit eastward to Manakot and the Karnali river, returning to Chainpur via Talkot. In June we travelled west to the Kali Gad and on to the Marma district on the upper Chamlia river. Thence we went to Baitadi, and finished at Pithoragarh in Kumaon.

I then went off to collect in Kashmir.

The second journey in Nepal was made between 12 September and 3 October. Starting from Trisuli in Central Nepal I went north-west to Satsae khola, and thence via Gatlang to Langtang. I returned to Kathmandu via the Trisuli valley.

In this year I made 264 gatherings of plants from Nepal.

1966

In this year I made two separate journeys.

Between 8–21 March I collected in the Rapti valley to the south of Kathmandu with T. B. Shrestha and M. L. Banerji, visiting Hetaura, Amlekhganj, and Narayangarh.

I then went off to collect in Sikkim.

Between 16 May and 1 August I collected with T. B. Shrestha in West Nepal. Starting from Pokhara we travelled via Dhorpatan, the Jang La, and Tibrikot to Jumla and the Rara lake. Thence we turned eastward to Maharigaon and Kaigaon. From here we descended the Bheri in July to Gotam, went round the south side of Hiunchuli Patan, and then crossed northwards over the Toridwari Bhanjyang to reach the lake at Ringmo. We then went by way of Sya Gompa and Tarap to Ccharka, crossed to Tukucha, and returned to Pokhara.

In this year I made 312 gatherings of plants from Nepal.

1967

This year I made six separate journeys.

Between 5–21 February I collected in Central Nepal with M. L. Banerji. Starting at Godavari at the south end of the Nepal valley we went to Hariharpur Garhi and Dungrebas, and thence to Chisapani on the Kamla khola. We finished at Janakpur.

Between 5–14 March I collected in East Nepal. Starting from Dharan I went west to the confluence of the Kosi, Arun, and Tamur rivers. I then went east along the Mahabharat lekhs and descended to the plains at Dangi, returning thence to Dharan.

Between 29 March and 28 May I collected in East and Central Nepal. Starting from Bhadrapur in the extreme east I went via Ilam, Chyangtapu, Yampodin, and Hellok to Taplejung. Thence I went to Chainpur, crossed the Arun at Num and reached Chaunrikharka on the Dudh Kosi via Cchoyang and the Salpa Bhanjyang. From here I went to Kathmandu by the main route.

I returned to Nepal in the autumn of this year with L. H. J. Williams. Between 21–28 August we collected in the Rapti valley of Central Nepal visiting Hetaura, Amlekhganj, and Narayangarh.

Between 1–23 September Williams and I collected in East Nepal. Starting from Dharan we travelled to Dhankuta, and went northwards along the ridge to the Milke Dara. Thence we went to Terhathum and crossed the Tamur to reach Rakshi Dara on the Mahabharat range. From here we descended into the bhahar and returned to Dharan.

On these two trips 459 gatherings of plants were made under joint Williams and Stainton numbers.

Between 27 September and 29 October I collected in Central Nepal. From Kathmandu I went via Trisuli and Satsae khola to cross the Buri Gandaki at Khorlak. Thence I went to Barpak and up towards the Rupina La. From here I went to Sisaghat, then north to Lamjung Himal, and finished at Pokhara.

On this latter trip and the ones made in the earlier part of the year I made 490 gatherings.

1968

In this year I made two separate journeys, both in West Nepal.

Between 7 March and 12 April I travelled from Nepalganj to Surkhet, and up the Karnali valley to Raskot. I then turned east across the Punge lekh into the Tila khola and continued south-east to cross the Sam La into the Sama khola. Thence I went to Jajarkot, and returned to Nepalganj via the Bheri valley.

From 26 April to 17 July I travelled from Pokhara to Simikot in the far north-west and back again. First I made a diversion of ten days to collect on Lamjung Himal, and then I travelled from Pokhara to Dhorpatan and along the southern side of the main range to cross the Bheri at Gotam. Thence I went via Hurta, Munigaon, and the Bundi Lagna to the Rara lake. I went west down the Khater khola to cross the Karnali, and then north over the Munya pass to Simikot. From here I returned over the Chankeli pass and turned east to Mugu. I then crossed southwards over Sisne Himal to Maharigaon, and went to Kaig-aon, Tibrikot, and Tarakot. From here I turned south over the Jang La to Dhorpatan, and returned to Pokhara.

In this year I made 255 gatherings of plants.

1969

In this year I made two separate journeys.

Between 15 February and 28 March I collected in East Nepal. Starting from Dharan I went eastwards along the Mahabharat lekh and descended into the Mai khola. I then continued eastwards to Sanichari and the Mechi khola. From here I followed the west side of the Singalelah ridge northwards to Sandakphu, and then turned south to Ilam and Soktim. Crossing westward over the Mai khola I returned to Dharan.

Between 19 September and 9 November I collected in Central and East Nepal. Leaving Kathmandu I travelled by the main route to Khumbu, Thyangboche, and the Everest basecamp. From Khumbu I turned south to Aisyalukharka, Halesi and Udaipur Garhi, and thence eastwards down the Trijuga khola to Dharan.

In this year I made 285 gatherings of plants.

References

INTRODUCTION pp. 1–4

1. For an account of these expeditions see Tilman, H. W. (1952), *Nepal Himalaya*. Cambridge University Press.
2. For accounts of these expeditions see:
   Williams, L. H. J. (1953), 'The 1952 expedition to Western Nepal', *J. Roy. Hort. Soc.*, vol. 78, pt 9, p. 323.
   Sykes, W. (1955/6), 'The 1954 expedition to Nepal', ibid., vols. 80 and 81, pts 1 and 2, p. 538.
3. In the three years which have elapsed since writing this J. F. Dobremez of Grenoble University, a trained ecologist, has taken to the field on behalf of the Centre National de la Recherche Cooperative with the object of making an ecological map of Nepal. The first sheet, which covers the Annapurna–Dhaulagiri region on a scale of 1/250,000, was published in 1971, and the whole project when completed will provide a very important addition to our knowledge of the country.
4. Hara, H. (1963), *Spring Flora of Sikkim Himalaya*. Hoikusha, Japan.
   —— (1968), *Photo-album of Plants of Eastern Himalaya*. Inoue Book Co., Tokyo.
   —— (1966), *The Flora of Eastern Himalaya*. University of Tokyo.
   Nakao, S. (1964), *Living Himalayan Flowers*. Mainichi Newspapers, Tokyo.

CLIMATE pp. 5–13

1. Mason, K. (1955), *Abode of Snow*, p. 46. Hart-Davies, London.
2. Dobby, E. H. G. (1966), *Monsoon Asia*, 3rd ed., p. 231. Univ. London Press.
3. Mason, K. (1936), 'Rainfall and rainy days in the Himalaya west of Nepal', *Himalayan J.*, vol. 8, p. 92.
4. Hagen, T. (1963), *Mt Everest*, p. 81. Oxford Univ. Press.
5. Hooker, J. D. and Thomsom, T. (1855), *Flora Indica*, introductory essay, p. 172. London.

CLIMATIC AND VEGETATIONAL DIVISIONS OF NEPAL pp. 15–32

1. Burkhill, J. H. (1910), 'Notes from a journey to Nepal', *Rec. Bot. Surv. India*, IV, p. 59.
2. Schmid, E. (1938), 'Contribution to the knowledge of flora and vegetation in the Central Himalayas', *Journ. Ind. Bot. Soc.*, XVII, 4, p. 269.
3. Kawakita, J. (1956), *Land and crops of Nepal Himalaya*, p. 42–52., Kyoto.
4. Schmid, E., ibid., p. 272.
5. Kawakita, J. (1956), ibid., p. 42.

FOREST TYPES pp. 53–135

1. Schweinfurth, U. (1957), *Die horizontale und vertikale Verbreitung der Vegetation im Himalaya*. Bonn.

REFERENCES

2. Osmaston, A. E. (1927), *A Forest Flora for Kumaon*. Allahabad.
3. Champion, H. G. (1936). 'A preliminary survey of the forest types of India and Burma.' *Indian Forest Rec.* (New series), Silvicult., I.
4. Choudhury, K. C. R. (1951), 'Sikkim—the country and its forests', *Indian Forester*, 77, p. 676.
5. Cowan, J. M. (1929), 'The forests of Kalimpong; an ecological account', *Rec. Bot. Surv. India*, XII, No. 1, p. 27.
6. Osmaston, A. E. (1927), *A Forest Flora for Kumaon*, p. xv. Allahabad.
7. King, G. (1889), 'The Indo-Malayan species of Quercus and Castanopsis', *Ann. Roy. Bot. Gard.* (*Calcutta*), at p. 26.
8. Hara, H. (1966), *The Flora of Eastern Himalaya*. University of Tokyo, at p. 50.
9. Bor, N. L. (1938), 'A sketch of the vegetation of the Aka hills, Assam', *Indian Forest Rec.*, vol. 1, No. 4.
10. Biswas, K. (1940), 'The flora of the Aka hills', *Indian Forest Rec.* (*New Series*) *Botany*, vol. 3, No. 1.
11. Kawakita, J. (1956), *Land and Crops of Nepal Himalaya*, p. 51. Kyoto.
12. Choudhury, K. C. R. (1951), 'Sikkim—the country and its forests', *Indian Forester*, vol. 77, No. 11, p. 676.
13. King, G. (1889), 'The Indo-Malayan species of Quercus and Castanopsis', *Ann. Roy. Bot. Gard.* (*Calcutta*), at p. 32.
14. Kanai, H. (1966), *The Flora of Eastern Himalaya*, at p. 51. Tokyo University.
15. Grubb, P. J. (1971), *Nature*, vol. 229, p. 44, Jan.
16. Choudhury, K. C. R. (1951), 'Sikkim—the country and its forests', *Indian Forester*, vol. 77, at p. 681.
17. Kawakita, J. (1956), *Land and Crops of Nepal Himalaya*, p. 43–50. Kyoto.
18. Do Amaral Franco, J. (1969), 'On Himalayan-Chinese cypresses', *Portugaliae Acta Biol.*, vol. 9, No. 3–4. p. 183.
19. Bor, N. L. (1938), 'A sketch of the vegetation of the Aka hills, Assam', *Indian Forest Rec.*, vol. 1, No. 4, at p. 187.
20. Kawakita, J. (1956), *Land and Crops of Nepal Himalaya*, p. 48. Kyoto.
21. Shebbeare, E. O. (1934), 'The conifers of the Sikkim Himalaya and adjoining country', *Indian Forester*, p. 710.
22. Kawakita, J., ibid., p. 51.
23. Kitamura, S. (1955), *Fauna and Flora of Nepal Himalaya*, at p. 82. Kyoto.

NOTES ON DISTRIBUTION pp. 137–169
1. Puri, G. S. (1960), *Indian Forest Ecology*, vol. 1, p. 62. Oxford Book and Stationery Co., New Delhi.
2. Hara, H. (1966), *The Flora of Eastern Himalaya*, p. 188. Tokyo.
3. Hooker, J. D., & Thomson, T. (1855), Introductory essay to *Flora Indica*, at p. 112. London.
4. Ludlow, F. (1944), 'Birds of S-E. Tibet', *Ibis*, Jan., p. 61.
5. Ward, F. K. (1936). *Proceedings of the Linn. Soc.*, The Hooker Lecture, at p. 142.
6. Choudhury, K. C. R. (1951), 'Sikkim—the country and its forests', *Indian Forester*, vol. 77, No. 11, pp. 676 and 681.

7. Ward, F. K. (1938), 'The Assam Himalaya', *J. R. Cent. Asian Soc.*, p. 610.

8. Ludlow, F. (1937), 'The birds of Bhutan and adjacent territories of Sikkim and Tibet', *Ibis*, Jan., p. 5.

9. Since writing the above comments on Bhutan my attention has been drawn to an article written by D. B. Deb, G. Sen Gupta, and K. C. Malik entitled 'A contribution to the flora of Bhutan' in the *Bull. Bot. Soc. Bengal* (1968), vol. 22, No. 2. Their description on pp. 173–174 of the climate of Bhutan reinforces my view that some parts of the country are dry, and that species that are absent from Sikkim are able to survive in these drier parts.

10. Airy-Shaw, H. K. (1934), 'A key to the species of Deutzia', *Kew Bull.*, p. 181.

11. In a recent article in *Plant Life of South-West Asia* (Davis, Harper, & Hedge, Bot. Soc. of Edinburgh, 1971, at p. 53), entitled 'Mediterranean elements in the flora and vegetation of the Western Himalayas', H. Meusel points out the connection between certain elements in the floras of Western China and the Western Himalaya, and the further connection of both floras with those of South-West Asia and the Mediterranean. This article is very relevant to many of the problems discussed above, and makes one realise that questions of Himalayan distribution cannot be solved exclusively in Sino–Himalayan terms.

GEOGRAPHICAL INDEX

To find the description of any forest type turn to the list on pp. 56–57.
References to species are not indexed.